Tolley's Risk Assessment Workbook Ser
OFFICES

by

Edward G Hines FRSH MIOSH RSP
Principal Consultant, Sypol Ltd

Series Editor: Allan St John Holt FIOSH RSP
Head of Safety, Royal Mail Group

Members of the LexisNexis Group worldwide

United Kingdom	LexisNexis UK, a Division of Reed Elsevier (UK) Ltd, 2 Addiscombe Road, CROYDON CR9 5AF
Argentina	LexisNexis Argentina, BUENOS AIRES
Australia	LexisNexis Butterworths, CHATSWOOD, New South Wales
Austria	LexisNexis Verlag ARD Orac GmbH & Co KG, VIENNA
Canada	LexisNexis Butterworths, MARKHAM, Ontario
Chile	LexisNexis Chile Ltda, SANTIAGO DE CHILE
Czech Republic	Nakladatelství Orac sro, PRAGUE
France	Editions du Juris-Classeur SA, PARIS
Germany	LexisNexis Deutschland GmbH, FRANKFURT and MUNSTER
Hong Kong	LexisNexis Butterworths, HONG KONG
Hungary	HVG-Orac, BUDAPEST
India	LexisNexis Butterworths, NEW DELHI
Ireland	Butterworths (Ireland) Ltd, DUBLIN
Italy	Giuffrè Editore, MILAN
Malaysia	Malayan Law Journal Sdn Bhd, KUALA LUMPUR
New Zealand	LexisNexis Butterworths, WELLINGTON
Poland	Wydawnictwo Prawnicze LexisNexis, WARSAW
Singapore	LexisNexis Butterworths, SINGAPORE
South Africa	LexisNexis Butterworths, Durban
Switzerland	Stämpfli Verlag AG, BERNE
USA	LexisNexis, DAYTON, Ohio

First published in 2003
Reprinted 2003

A CIP Catalogue record for this book is available from the British Library.

ISBN 0 7545 1889 2

Typeset by Letterpart Limited, Reigate, Surrey
Printed and bound in Great Britain by Antony Rowe Ltd, Chippenham, Wilts
Visit LexisNexis UK at www.lexisnexis.co.uk

Preface

Risk assessment is carried out in order to evaluate the extent and significance of workplace hazards so that appropriate control measures can be put in place. Usually, a written record of the assessment is required. The busy manager who has to start with a blank sheet of paper faces an uphill struggle – most of us have been in that position at some time. Failure to carry out risk assessments (an issue increasingly coming before the courts) is often due not to lack of will but to lack of a suitable system. This series of Workbooks aims to provide the system, and to provide templates with some of the work already done.

There are many ways of assessing risks, some difficult and some easy to follow. There is now a trend towards task-based assessments, which can be more difficult to carry out. Whichever method is chosen, it is important that the assessments are tailored to the particular circumstances in the workplace so as to evaluate real risks and not theoretical ones. The work done by the authors of the Workbooks is therefore the basis for an assessment system, which needs to be completed in detail by the user.

The Workbooks offer an assessment method that ensures compliance with the law, and a practical system that can be understood easily, put into place quickly where necessary and offers real benefits to the users especially by encouraging updating and review. The rules are fully explained, although the principle is simple. Perhaps the simplest and best summary of what needs to be done was published in the Health and Safety Executive's leaflet 'Five Steps to Risk Assessment' in 1994. Any employer who adheres to these guidelines will comply with the law, and perhaps more importantly, create a safer workplace.

Allan St John Holt

Acknowledgements

The Publishers would like to kindly thank Sypol for allowing their generic risk assessment templates to be adapted for use in the Workbooks.

About the Author

Edward G Hines

Edward G Hines is the Principal Health and Safety Consultant at Sypol Ltd. His experience and academic qualifications span amongst others financial institutions, fire safety and engineering, the rail industry, commercial premises and warehousing, NHS and private healthcare, residential and nursing homes, local authorities, charities and airline operations.

Edward's practical safety qualifications include being an accredited NEBOSH trainer, as well as a Registered Safety Practitioner, a Member of the Institution of Occupational Safety and Health and a Fellow of the Royal Society of Health. He has authored, edited and reviewed a wide range of health and safety publications and technical papers.

Series Editor

Allan St John Holt

Allan St John Holt is one of Britain's best-known safety experts. He has also lectured widely in the USA, the Far East and Australia on health and safety matters. Twice President of the Institution of Occupational Safety and Health (IOSH), and a Fellow of the Institution, he is a Registered Safety Practitioner. He is currently Head of Safety for the Royal Mail Group.

A social studies graduate, Allan founded the National Examination Board in Occupational Safety and Health in 1979. An Ambassador and member of the Board of Directors of Veterans of Safety International, he was inducted into the Safety and Health Hall of Fame International in October 1997, and was elected Honorary President of the Southampton Occupational Safety Association in 1991.

Contents

Part 1

Risk Assessment

Risk Assessment in the Office

Part 2

Checklists, Questionnaires and Action Plans

Part 1

Risk Assessment

Introduction

Risk assessment is not a new idea. The principle of relating the amount of action required to the size of the problem was set out almost 30 years ago within the *Health and Safety at Work etc Act 1974.* This was the first major piece of health and safety law to avoid giving directions and rely instead upon a management evaluation of what is needed to take reasonable care of the health and safety of both employees and anyone else affected by the work activities of an organisation. The employer could only do what is 'reasonably practicable' by evaluating the hazards and risks involved and applying appropriate controls.

The first health and safety law to require a specific evaluation, or risk assessment, was the *Control of Lead at Work Regulations 1980 (SI 1980 No 1248).* This was followed seven years later by Regulations controlling work with asbestos (*Control of Asbestos at Work Regulations 1987 (SI 1987 No 2115)*) and then by the *Control of Substances Hazardous to Health Regulations 1988 (SI 1988 No 1657) (COSHH Regulations)* which came into force in October 1989. In 1993 the set of Regulations based upon new European Council Directives on health and safety was based on the principle of risk assessment. In many ways, the most important of these was the *Management of Health and Safety at Work Regulations 1999 (1999 No 3242) (MHSWR 1999),* introducing risk assessment as a requirement for all work situations other than those already covered by more specific assessment requirements.

The Regulations, although now revised and expanded from their original form, do not specify any particular method of carrying out risk assessments and, as a result, there are many different systems in use. Experience shows that the process does not have to be especially time consuming provided that a format is used that presents and records information in a clear and logical way. One of the few direct requirements of all the Regulations on the subject is that the results must be recorded in writing where there are five or more employees, so the employer's chosen format is important in order to show that his/her assessments are suitable and sufficient.

Risk assessment is the key to successful management of health and safety at work. The law no longer provides a list of 'what to do': it now recognises that life is more complicated than that. Employers must work out for themselves what the problems are:

- the hazards – and how much of a problem each one is;
- the risk – and then work out appropriate control measures.

This is the essence of risk assessment. It is a continuous process, and a full explanation of it in the context of the systematic management of health and safety can be found in the Health & Safety Executive's (HSE) publication HSG65 'Successful Health and Safety Management'.

Once the assessment is completed and recorded, the document becomes a blueprint for action to remove hazards and lower the risks associated with each work activity. Only then can the employer establish policies to control the risks he is now aware of, and set up procedures. Risk assessments also need to be reviewed, as conditions change over time. *Management of Health and Safety at Work Regulations 1999 (1999 No 3242) (MHSWR 1999)* and specific legislation specifies the need to review; this must be done:

- when new materials and equipment are introduced;
- following significant changes in work practices and premises, and when monitoring shows that control measures are not working properly.

Following accidents, the assessments should be reviewed for accuracy, to make sure that the circumstances have been taken fully into account.

The key is to identify significant hazards and risks. In some cases, where the situation changes rapidly because of the nature of the work, risk assessment will identify a range of hazards and their attendant risks and specify general controls, relying upon selection of employees and appropriate training to enable them to take control of specific risks as they arise.

Risk assessments can be made of premises, machines, processes and tasks. In some circumstances, it has been found easiest to make assessments that are task or process based – common examples include driving company vehicles, using fork lift trucks, loading and unloading goods.

Terminology

Some definitions

The words 'hazard' and 'risk' have special meanings in health and safety, which are narrower and more specific then their customary definitions.

Hazard means the inherent property or ability of something or some situation to cause harm, loss or damage. A hazard has the potential to interrupt or interfere with a process or a person. **Risk** is the chance, likelihood or probability of the hazard actually causing the loss – an evaluation of the potential within the hazard.

So a simple way to remember the difference is that 'hazard' describes potential for harm, and 'risk' is the chance that harm will result in the particular circumstances.

Risk has two main components, the likelihood of the failure occurring, and the severity of the consequences of failure. Comparisons between risks can be made by giving numerical values to these factors. This is helpful in allowing an estimate to be made of the reduction in the risk that can be made by the introduction of controls, and also in establishing priorities for action.

It is important to be aware of the limitations in the use of numerical risk ratings. For example, the probability of failure and the extent of the outcome cannot be predicted for individuals in many cases of occupational ill-health exposure. As a result, numerical scoring of risk is unreliable for evaluating the extent of occupational health hazards. Some alternative methods of carrying out and recording risk assessments use a simple classification such as Insignificant, Low, Medium and High in order to avoid any confusion or argument about what numbers might mean in any particular circumstance.

Risk assessments are required to be 'suitable' and 'sufficient' for the purpose. Such an assessment will reflect what it is reasonably practicable for the assessor to know or find out about the hazards in the workplace. Therefore, the assessor must be **competent**, and this means someone with an appropriate combination of knowledge, training and experience in the industry to enable them to:

- evaluate the impact of specific legal requirements and the need for risk assessments;
- formulate an appropriate action plan to remove the hazards and control the risks;
- co-ordinate the implementation work.

Another part of *MHSWR 1999* requires management to appoint one or more competent persons, normally from amongst the employees, to assist the employer in complying with health and safety law. This person should either carry out or oversee the risk assessment process. Ability to do so will be one of the key competencies of the person so appointed.

(The section on RISK ASSESSMENT IN THE OFFICE gives details of specific Regulations applying to offices.)

Control measures

Some of the controls that can be used work better than others, and some can be very ineffective. In order of effectiveness, these are the accepted headings of controls available:

- **Hazard elimination** – the best solution of all is to remove the hazard, use of alternatives, design improvements, or just deciding not to attempt the work.
- **Substitution** – replacing a hazard can lower the risk, such as when replacing one chemical with another less dangerous one, or using a more appropriate (safer) piece of equipment.

- **Use of barriers – isolation**, such as putting the hazard in a box, for example the physical isolation of noisy equipment, and **segregation** (removing employees from the hazard by putting them in a box such as a protected cabin).
- **Use of procedures** – procedures can limit exposure time, safe systems of work can define acceptable limitations on the work – but these depend upon people to carry them out properly and they are less effective as a result – they can be easily avoided or ignored.
- **Use of warning systems** – signs, instructions, labels – all of which depend on human response and again can be ignored.
- **Use of PPE** – this depends totally on human response to be effective, assuming an appropriate piece of equipment has been selected. This is to be used as a sole measure only when all other options have been exhausted – personal protective equipment (PPE) is the last resort.

Putting it all together

Hazards and risks are sometimes difficult to identify and separate, so it will be helpful at this stage to think about a common activity that everyone can relate to and then use the Workbook format to analyse and record the assessment.

Most businesses use motor vehicles and so the activity of driving vehicles at work needs to be assessed. First, we can list the main or significant **hazards** associated with motoring:

- collision with other vehicles and roadway structures;
- collision with permanent structures in the premises;
- collision with pedestrians in the premises;
- objects falling from the vehicle;
- vehicle defects;
- fire;

and there may be others, such as carriage of dangerous goods, specific to the business.

The **risks** associated with these hazards will vary considerably with the particular circumstances, but some general statements can be made about them that will be valid and useful. For example, fire is relatively uncommon, and few people are involved in vehicle fires. On the whole, we can say that the fire risk is low, but the severity can be high. If flammable substances are carried in the vehicle the risk level would rise, and that shows **why risk assessments must always be related to the particular circumstances of the work and not just made for the general case**.

Other factors affecting risk will vary according to the individual circumstances. We have no influence on most of them, for example:

- stopping distance (speed and size, condition of vehicle, road conditions);
- speed and size of other vehicles;
- skill of drivers – which is related to age, experience, reaction times, physical condition;
- provision of safety devices for occupants;
- environmental conditions;
- traffic conditions;
- evasive skills of pedestrians.

What **controls** are already in place? A convenient classification for them is:

- Statutory – regulations concerning construction and use, licensing of drivers, requiring vehicle testing and speed restrictions.
- Guidance – Highway Code and other information resources.
- Planning – daily and weekly checks on the vehicle, checking the route before starting out and checking on the condition of the driver.
- Physical – measures include pedestrian crossing points, roadside barriers, signposts, seatbelts.
- Monitoring – carried out by the police, roadside cameras, spot checks on vehicles, and the monitoring generally of accidents and traffic patterns.

- Training – driver training is controlled and standards are laid down. Training of vulnerable groups of pedestrians (such as school children) is done using the Highway Code but is not a legal requirement.

Therefore, all these measures are already in place. The task of the risk assessment is to look beyond the general to the particular. This will mean assembling information, especially previous incidents. The list of hazards can be revisited and related specifically to the work being done, and can be set out as follows.

Basic risk assessment

Hazard	Risk level	Who is at risk?	Additional controls needed	Action by:
Collision with other vehicles and roadway structures	Low – low vehicle mileage, no significant accident record	Drivers, third parties	None	
Collision with permanent structures in the premises	Low – some structural damage recorded	Drivers	Improved road markings, road signs	Within 28 days
Collision with pedestrians in the premises	High	Third parties	Restrict access, separate pedestrians from roads	Within 7 days
Objects falling from the vehicle	Low – no unusual or unsecured loads carried	Third parties	None	
Vehicle defects	Low	Drivers	Planned maintenance system to be reviewed	Within 28 days
Fire	Medium – occasional flammable loads carried short distances within the premises	Drivers	Supply vehicle fire extinguishers to identified vehicles	Within 7 days

Rating the risks

Quantitative risk assessments produce a probability estimate based upon known risk information applied to the circumstances being considered. They can be found in 'safety cases' in high risk industries where, for example, they might indicate the probability of a nuclear incident or serious chemical leak.

Qualitative assessments are subjective, based on the personal judgement of the assessor, backed by more general information about probability. These are much simpler to make, and are those normally referred to by legal requirements. We know this because the HSE published guidance in 1994 entitled 'Five Steps to Risk Assessment', a leaflet setting out the most basic steps in the process and providing a specimen format for recording the findings of assessments. The leaflet has since been updated and reissued. Comments contained in the text give useful pointers to the approach to risk assessment which will satisfy minimum requirements in the view of the enforcing authority. For example:

> 'An assessment of risk is nothing more than a careful examination of what, in your work, could cause harm to people so that you can weigh up whether you have taken enough precautions or should do more . . .'.

We are told there is no need to make assessments complicated. Nevertheless, there are benefits to be gained by estimating the level of risk associated with hazards, in particular because we can then make an evaluation of the expected benefits of the control measures contained in the assessment – a before and after evaluation. The choice of whether to do this numerically, or descriptively, or at all, is left to the individual.

A numerical approach has its drawbacks. Two or more variables can be used to produce a final rating by multiplication, as shown in the following Table. But there should be a method of ensuring that any high-scoring severity evaluation forces the problem to be studied at an early stage – even if the multiplication exercise puts it low in the priority list. For example, an event with a high probability of occurrence but a low severity rate could be rated the same as an event with a low probability of occurrence but a high severity rate. The numbers themselves have no inherent value – a risk rated at 20 is not 'half as dangerous' as one rated at 10. The purpose of the exercise is simply to allow priorities to be established and to demonstrate the effectiveness of the controls put in place.

A descriptive approach simply uses words such as 'insignificant', 'low', 'medium' and 'high' in their normal sense to evaluate the product of the likely severity of the outcome and the probability of its occurrence. Most people have an appreciation of what 'Medium Risk' means for them in their business context, and can appreciate that with controls in place the risk should be lower – preferable to 'Insignificant' or at the least to 'Low'.

Numerical Assessment Table

Probability (likelihood of the event)	
Almost certain	5
Probable	4
Could happen occasionally	3
Improbable	2
Almost impossible to happen	1

Severity (the most likely result of the accident)	
Catastrophic – death	5
Severe incapacity	4
Absent one month	3
Absent three days + recovery	2
First aid treatment required only	1

Assess the risk, multiply the two numbers and transfer to the **Risk Index Table**:

		Probability				
Severity		5	4	3	2	1
	5	25	20	15	10	5
	4	20	16	12	8	4
	3	15	12	9	6	3
	2	10	8	6	4	2
	1	5	4	3	2	1

Typical actions required

Priority 1	25–16	Extensive precautions based on physical controls, work method statements, specific and stringent rules with significant penalties for infringement.
Priority 2	15–10	Training, checks on competence of employees needed, high level of supervision.
Priority 3	9–1	Some risk acceptable, competent employees appropriate level of supervision.

Overview of the law regarding risk assessment

Health and Safety at Work etc Act 1974

This Act is the central piece of health and safety legislation in Great Britain. The bulk of the Act is now historical, as it 'enabled' the mechanisms for establishing the Health and Safety Commission and Executive, the Medical Advisory Service (EMAS) and so forth. The substantive portions are contained in the General Duties sections of the *Health and Safety at Work etc Act 1974 (HSWA 1974), ss 2–9.*

General duties of employers are set out in *sections 2–4,* and *9.* As far as is reasonably practicable, employers must safeguard the health, safety and welfare of their employees and anyone else who may be affected by the work activities. This duty extends to providing and maintaining:

- safe plant and systems of work;
- safe handling, storage, maintenance and transport of work articles and substances;
- necessary information, instruction, training and supervision;
- a safe place of work and means of getting to and from it safely;
- a safe working environment with adequate welfare facilities.

The details are left to the employer to work out, but by 1992 a set of six (The Six Pack) more specific pieces of safety law was added to cover both general issues (management, safe workplace and equipment) and specific ones such as personal protective equipment and manual handling. The requirement to make risk assessments was not a feature of the Act, but it can be seen that in order to discharge the general duties, it will be necessary for employers to make an evaluation of health and safety issues facing them. The specific risk assessment requirement for all employers is contained in the *Management of Health and Safety at Work Regulations 1999 (SI 1999 No 3242)*, which first appeared in 1992.

Management of Health and Safety at Work Regulations 1999

The *Management of Health and Safety at Work Regulations 1999 (SI 1999 No 3242) (MHSWR 1999)* provide most of the detail lacking in the *Health and Safety at Work etc Act 1974*, and importantly specify the requirements for risk assessment of all work activities. It is strongly advised to consult a copy of the accompanying Approved Code of Practice (ACoP) which contains both the Regulations and an authoritative interpretation. Employers should be familiar with all of the Regulations, but in relation to risk assessment the following are particularly relevant.

Regulation 3

Under *MHSWR 1999 (SI 1999 No 3242), reg 3*, every employer and self-employed person is to make a suitable and sufficient assessment of the health and safety risks to his employees, and others not in his employment to which his undertakings give rise, in order to put in place appropriate control measures. It requires appropriate review to be made of the assessments, and for the significant findings to be recorded if five or more are employed. Details are also to be recorded of any group of employees identified by an assessment as being especially at risk.

Regulation 3 also requires employers to make a specific risk assessment where 'young persons' are employed (these are people under the age of 18). In making or reviewing the assessments, the following must be taken into account:

- the inexperience, immaturity and lack of awareness of risks of young people;
- the fitting-out and layout the of the workplace;
- nature, degree and duration of exposure to physical, chemical and biological agents;
- type, range and use of work equipment;
- organisation of processes and activities;
- risks from special processes which are listed.

A similar provision covering specific risk assessment of new and expectant mothers is to be found in *Regulation 16.*

Regulation 5

MHSWR 1999 (SI 1999 No 3242), reg 5 requires the employer to make appropriate arrangements, given the nature and size of his operations, for effective planning, organising, control, monitoring and review of the preventive and protective measures he puts in place. If there are more than five employees the arrangements are to be recorded. (The measures themselves will have been derived from the risk assessments.)

Regulation 6

MHSWR 1999 (SI 1999 No 3242), reg 6 requires the employer to provide appropriate health surveillance, identified as being necessary by the risk assessments.

Regulation 7

MHSWR 1999 (SI 1999 No 3242), reg 7 deals with the appointment of adequate numbers of competent persons to assist the employer to comply with obligations under all the health and safety legislation, unless (in the case only of sole traders or partnerships) the employer already has sufficient competence to comply without assistance. *Regulation* 7 and the ACoP set out clearly what is required of the competent person and the appointment arrangements.

Regulation 8

MHSWR 1999 (SI 1999 No 3242), reg 8 requires employers to establish and give effect to procedures to be followed in the event of 'serious and imminent danger' to persons working in their undertakings, to nominate competent persons to implement any evacuation procedures and restrict access to danger areas. The identification of potential 'serious and imminent danger' is one of the functions of the risk assessment process.

Regulation 10

MHSWR 1999 (SI 1999 No 3242), reg 10 covers the provision of information by the employer on health and safety risks identified by the assessments, plus all the control measures, together with information on the identity of competent persons appointed under *Regulation 8*. There is also more on the employment of children and young persons – before employing them, the employer must provide the parents with comprehensible and relevant information on assessed risks and the control measures in place.

Specific risk assessment legislation summaries

Control of Substances Hazardous to Health Regulations 2002

First introduced in 1988, the *Control of Substances Hazardous to Health Regulations 2002 (SI 2002 No 2677) (COSHH Regulations)* gave the first general requirement for risk assessment, as almost all employers have, make or use substances falling within scope. Although modified and expanded since the original version, the essence of *COSHH* remains the same: prevention of workplace disease and ill health resulting from exposure to hazardous substances. Again, there is a control framework beginning with the assessment of risks to health arising from work activities associated with hazardous substances, introduction of control measures, maintenance of the measures and associated equipment, and monitoring their effectiveness.

Health and safety law also requires information to be made available by the supplier of substances; this provides a good starting point for carrying out risk assessments. Possession of a data sheet is not the same as having done a risk assessment, though. Under the *COSHH Regulations*, there is a duty on the employer to ensure that exposure to substances hazardous to health is either prevented, or where this is not reasonably practicable, controlled. Prevention must initially be attempted by means other than the provision and use of personal protective equipment by employees.

Any substance used at work, or deriving from work activities, is a 'COSHH substance' if it is harmful to people's health in the form in which it occurs at work. Most common substances meeting this definition will be labelled on the container as 'dangerous', using an orange diamond with a black symbol inside it to indicate the contents are toxic, very toxic, harmful or corrosive. However, substantial quantities of dust are also included in the *COSHH* definition, and there may be substances produced as by-products of the work processes that cannot be labelled in this way.

Personal Protective Equipment at Work Regulations 1992

The *Personal Protective Equipment at Work Regulations 1992 (SI 1992 No 1144), reg 4* places a duty on the employer to provide 'suitable personal protective equipment (PPE) to employees, except where risks have been adequately controlled by other equally, or more effective, means. It is to be used as a last resort, steps having been taken to remove hazards or control risks by means of safer processes, systems or conditions. The need for PPE will have been identified by risk assessment.

Regulation 6 requires the employer to make a specific risk assessment to determine whether the proposed PPE is in fact 'suitable' – appropriate for the risks involved and the conditions, capable of fitting the wearer properly after adjustment, and whether it will be effective in controlling the risk.

The assessment must include an evaluation of the risks that have not been avoided by any other means, and a comparison of the characteristics the PPE needs to have to be effective against the risks against the characteristics of the proposed PPE. For high-risk situations and/or where the PPE is complex, the assessment should be in writing and kept available.

Where PPE is to be used, the employer should be aware of other detailed requirements of these Regulations which relate to training and information, storage and maintenance of the PPE.

Manual Handling Operations Regulations 1992

The *Manual Handling Operations Regulations 1992 (SI 1992 No 2793)* focus attention on ergonomic solutions to manual handling issues, again based upon risk assessment. *Regulation 4* is of crucial importance, requiring employers to avoid the need for employees to carry out those manual handling operations that involve a risk of injury as far as is reasonably practicable. Where this cannot be done, a 'suitable and sufficient' risk assessment is required for all such operations, so as to take the appropriate steps to reduce the risk of injury to the lowest level reasonably practicable. The ACoP to the Regulations is particularly helpful as it contains a specimen assessment form and worked examples.

Assessments are required to cover specific questions, which are contained in the Schedule to the Regulations, and to some extent this prescribes the format of the assessment to be used.

Health and Safety (Display Screen Equipment) Regulations 1992

The *Health and Safety (Display Screen Equipment) Regulations 1992 (SI 1992 No 2792)* are aimed at improving working conditions at display screen equipment by providing ergonomic solutions, as well as to enable users of the equipment to obtain information about the hazards, risks and their control measures together with eye and eyesight tests. The specific risk assessment requirement is found in *Regulation 2*, where employers are required to carry out a suitable and sufficient analysis of each workstation, and to review following any significant changes. Major changes in software used and task requirements, hardware and work equipment would be classed as significant. Unusually, the Regulations contain a list of minimum requirements for workstations in the Schedule to the Regulations; this is derived from the EU Directive (*90/279/EEC*).

The need to comply

A failure to carry out risk assessments of an adequate kind renders the employer vulnerable to criminal prosecution and, in relation to specific requirements of Regulations, including the *Health and Safety (Display Screen Equipment) Regulations 1992 (SI 1992 No 2792)* and the *Manual Handling Operations Regulations 1992 (SI 1992 No 2793)*, to successful civil claims for compensation by injured employees.

The level of penalties applied so far by the courts for failure to make risk assessments is becoming significant. This offence is usually not the sole charge brought against an employer, but it is a charge added in an increasing proportion of cases. Put at its simplest, the employer without any or any adequate, suitable and sufficient risk assessments cannot show that the most basic of tasks required of him has been completed – that of having an effective system for managing health and safety at work. Thus he is vulnerable to charges under the Act, as well as several Regulations summarised here.

The good news is that the process is not normally complicated, or one that requires special training beyond an appreciation of what is being done and a knowledge of the particular business.

This workbook supplies the means for basic risk assessment in offices, giving templates and worked examples of common hazards and their usual control measures. Details of applicable legislation can be found in the section on RISK ASSESSMENT IN THE OFFICE.

Table of legislation

- *Noise at Work Regulations 1989 (SI 1989 No 1790)*;
- *Management of Health and Safety at Work Regulations 1999 (SI 1999 No 3242);*
- *Health and Safety (Display Screen Equipment) Regulations 1992 (SI 1992 No 2792)*;
- *Workplace (Health, Safety and Welfare) Regulations 1992 (SI 1992 No 3004)*;
- *Provision and Use of Work Equipment Regulations 1998 (SI 1998 No 2306)*;
- *Personal Protective Equipment at Work Regulations 1992 (SI 1992 No 2966)*;
- *Health and Safety (Miscellaneous Amendments) Regulations 2002 (SI 2002 No 2174)*;
- *Control of Substances Hazardous to Health Regulations 2002 (SI 2002 No 2677)*;
- *Fire Certificates (Special Premises) Regulations 1976 (SI 1976 No 2003)*;
- *Fire Precautions (Application for a Certificate) Regulations 1989 (SI 1989 No 77)*;
- *Fire Precautions (Non-certificated Factory, Office, Shop and Railway Premises) (Revocations) Regulations 1989 (SI 1989 No 78)*;
- *Fire Precautions (Workplace) Regulations 1997 (SI 1997 No 1840) (as amended by SI 1999 No 1877)*;
- *Fire Safety and Safety of Places of Sport Act 1987.*

Guidance material

Health & Safety Executive:

- L22 – 'Safe Use of Work Equipment – Guidance on Regulations';
- L23 – 'Manual Handling – Guidance on Regulations';
- L25 – 'Personal Protective Equipment at Work – Guidance on Regulations';
- L26 – 'Display Screen Equipment – Guidance on Regulations';
- HSG53 – 'RPE – Selection, Use and Maintenance';
- HSG88 – 'Hand-arm Vibration';
- HSG121 – 'A pain in your workplace – ergonomic problems and solutions';
- HSG218 – 'Tackling work-related stress';
- INDG171 – 'Upper Limb Disorders – Assessing the Risk';
- INDG341 – 'Stress leaflet';
- INDG281 – 'Short guide to stress'.

Home Office:

- Code of Practice: 'Fire Precautions in Factories, Offices, Shops and Railway Premises not required to have a fire certificate';
- 'Fire Precautions Act 1971: Guide to fire precautions in existing places of work that require a fire certificate';
- 'Guide to fire precautions in existing places of entertainment and like places';
- 'Fire safety: an employer's guide', ISBN 0 11 341229 0.

How to use the workbook and methodology

Essential steps in risk assessment

The five basic requirements for successful risk assessment are:

- identify the hazards;
- decide who might be harmed (at risk), and how;
- evaluate the extent of the risk by looking at the action already being taken, then deciding if that is sufficient or if more needs to be done;
- recording the significant findings of the assessment;
- reviewing the assessment at appropriate intervals.

The Workbook provides a format to prompt and help identify the hazards, and to record the actions taken and those needed. Whether part of a review or a first risk assessment for any business, this action plan that will cover all the requirements if followed. The schematic approach summarised below can be used to explain to the assessment team the steps involved and their part in the work.

Action Plan

1. **Appoint one or more competent persons** to co-ordinate the risk assessment activities, with responsibility for checking the adequacy of current policy, arrangements and management systems for ensuring health and safety. The competent person should have access to the Approved Codes of Practice and guidance associated with the Regulations. Their task will be to evaluate the impact of the requirements and especially the need for risk assessments, carry out or supervise the carrying out of the risk assessments, put together a management plan to take the necessary actions and champion the actions. This workbook provides the means of making the required risk assessments and for sharing the results with employees (by training) and with other employers affected by the work activities.
2. **Assemble information.** Safety data sheets supplied by manufacturers of substances used at work, information on substances generated during work, makers' specifications on matters such as maintenance of equipment, plans and designs, workstation layouts, training records – depending upon the activities, all of the above and more may be helpful to the assessment process.
3. **Identify and list all the work activities**. Not all of them will be required to be assessed, if there are no hazards associated with them, or if the risks are insignificant.
4. **Identify all the hazards associated with each activity, substance, procedure and group of people.** It has been found that activity-based assessments are easiest to carry out, but there will be circumstances when individual pieces of equipment and locations will need to be assessed. The Workbook provides examples of these, and covers all the most common workplaces and work practices in the industry.
5. **Carry out risk assessments.** The templates in PART 2 of this Workbook have been specially designed to help with the process. Their use is explained in the section on RISK ASSESSMENT IN THE OFFICE. The basic principle is that they list the majority of hazards and controls likely to be appropriate in particular situations, and allow their assessment either numerically or descriptively. But use of the Workbook's examples alone is not sufficient: there are likely to be circumstances and hazards that they do not cover, and also the levels of risk associated with each case must be evaluated. Remember that an assessment which is both suitable and sufficient reflects what it is reasonably practicable to expect employers to know about the hazards specific to their workplaces. Evaluate the risk in workplaces by thinking about all relevant factors – including the numbers of people exposed, the duration of exposure, and the likely outcome of the risk if not controlled.
6. **Record basic control measures.** Write down the controls that are already in place, and also what extra needs to be done in order to lower the risk level as far as reasonably practicable. Assign actions and timescales to make sure the actions are carried out.
7. **Record the residual risk.** 'Residual risk' is the level of risk left after the controls have been put in place. Could injury still result? How many are exposed? How likely is it to happen with all the controls in place? To show the suitability and sufficiency of the risk assessment, the effectiveness of the control measures should be evaluated to produce a 'before and after' picture.
8. **Make sure the assessment is dated.** Also, give a date by which the assessment must be re-evaluated. Part of the accident investigation process will be to revisit the assessments to ensure they remain appropriate in the light of any new evidence, but they also need to be checked for adequacy in the light of current work practices and conditions.

Special cases – stress

Stress is the adverse reaction people have to excessive pressure or other types of demand placed upon them. Stress is, therefore, not 'good', or 'an illness'; the Industrial Injuries Board has ruled that stress-related illness at work is not to be classed as an industrial accident. At least one employee in five rate their job as 'very stressful'. Recent work carried out by the HSE shows that management standards for the control of stress need to be reviewed, and the risks assessed. There is a clear duty under the *Health and Safety at Work etc Act 1974* and the *MHSWR 1999* to do so. The same applies to the joined issues of violence and bullying at work.

The liability of the employer to pay compensation to an affected employee depends upon the facts of the particular case. There must always be:

- a duty of care owed to the employee;
- a failure to satisfy that duty;
- an injury or loss as a consequence.

The number of cases claiming compensation for stress-related illness has been increasing rapidly in recent years, but an Appeal Court ruling in February 2002 set out new guidelines for courts after setting aside three awards of £200,000 in total that had been made by lower courts to employees suffering from stress at work.

The Court of Appeal ruled that employees must raise concerns over stress at work with their employer before submitting a claim, in order to have any chance of gaining compensation. This means that employers can normally assume that their employees can withstand the pressures of the job unless they know of some particular problem or vulnerability and, where the employer offers a confidential counselling service with access to treatment, he will rarely be held in breach of the duty of care.

It follows that employees will need to raise their concerns regarding stress with their employers and give the employer the chance to do something about it. If they do not give the employer this opportunity they will find it hard to prove a breach of care. Equally, the employer will also have an obligation to provide the means of raising those concerns. The same arguments are likely to hold good for violence and bullying.

At the time of writing, the HSE's view is that the voluntary route to stress management at work should be explored before any legislation. A European Directive may be produced which would then force specific legislation to be introduced into the UK.

The HSE believes that management must strive to understand the issue, investigate absences from work and analyse them for cause. It advocates that risk assessment within a business should concentrate on seven major stressors (stress factors). These factors are:

1. the demands upon the individual – workload, exposure to hazards;
2. the extent of the control the individual possesses over the work to be done;
3. the level of support, i.e. training, peer support and individual factors;
4. relationships at work – bullying, harassment;
5. nature of the work role – ambiguity, conflicting roles;
6. the pressure for change within the business; and
7. the culture of the business – in relation to workload expectation, for example – is it the norm to work longer hours or take work home?

The following checklist can be used to assess the level of stress risk in a business or part of a business. Numerical values can be given to the answers to produce a score, but the aim should be to produce an action list which will address the deficiencies and thus reduce the risk level. The completed checklist should be dated, signed by the assessor(s), and kept on file.

Checklist

Demands upon the individual	**Yes/ No**	**Required actions**	**Person responsi-ble**	**Comple-tion date**
Does the recruitment process correctly match skills and ability requirements to the job?				
Is the allocation of work monitored?				
Is induction and job change training given and is refresher training available?				
Are the hours to be worked agreed on recruitment and realistic for the work required?				
Do staff understand the reasons for any temporary increases in work or production rates?				

Demands upon the individual	Yes/ No	Required actions	Person responsi- ble	Comple- tion date
Is there a monitoring system for working environment factors such as noise, vibration, humidity?				
Are the resources adequate to do the required work?				
Is there suitable and sufficient work equipment available for all jobs?				
Control over the work				
Is a job design process followed for all new jobs, including evaluation?				
Are the Working Time Regulations complied with and work patterns controlled?				
Do employees have an opportunity to plan their own work where possible?				
Is there an element of flexibility in task allocation and hours of work?				
Is there a leadership feedback process?				
Do management and staff meet together regularly to make decisions about the resolution of work issues?				
Support, training and other factors				
Are employees aware of the availability of medical and welfare support teams and how to get in touch with them?				
Is there a clear, open and transparent process for transfers and promotions?				
Is there a process for training and career development, and has it been explained to employees?				
Is there a process for assessing skills and competencies?				
Relationships				
Are staff made aware of the process to deal with harassment?				

Demands upon the individual	Yes/ No	Required actions	Person responsi- ble	Comple- tion date
Is there a grievance procedure in place?				
Is there an equal opportunities policy in place, which is complied with?				
Is there measurement of attendance and is an appropriate procedure used correctly?				
Is there a Code of Conduct for the business and is it used properly?				
Is there a system for reporting and controlling bullying, harrassment, drink or drug abuse?				
Role				
Is there an appraisal process, and is it being followed correctly?				
Are there clear job descriptions for each job?				
Are meetings with managers held regularly and the results recorded and shared with employees?				
Does the training process ensure that the correct training is available for all jobs?				
Change				
Are staff consulted and involved in changes that may affect them?				
Is 'change' risk assessed in terms of its likely impact on health and safety?				
Culture				
Is there any stress awareness training?				
Is there a process in place to deal with harassment and bullying?				
Is sick absence monitored regularly for signs of stress related illness?				
Is annual leave monitored to ensure that employees take their leave entitlements?				

Demands upon the individual	Yes/ No	Required actions	Person responsible	Completion date
Are staff discouraged from working excessive hours?				
Are reports of stress taken seriously and action taken?				
Are there good communication channels set up so that employees are kept well informed about the business?				
Is there a feedback process in place for staff comments and/or complaints?				

Special cases – fire

A fire risk assessment must be carried out for every workplace, regardless of the existence of any approval or certification which may have been given. It is a practical exercise, requiring a tour to verify its completeness and accuracy. The purposes of the assessment are to:

- identify the extent of the fire risk;
- assess the likelihood of a fire occurring;
- identify any additional precautions that may be needed, or control systems that do not function adequately.

The nature of the risk assessment is essentially the same – the tasks are:

- identify potential fire hazards;
- decide who might be harmed or at significant risk in the event of a fire, and identify their location;
- evaluate the levels of risk from each hazard, and decide whether existing precautions are adequate – or whether the hazard can be removed or controlled more effectively;
- record the findings and advise employees of the results, prepare or revise emergency plan;
- review the assessment as necessary, as conditions change and at regular intervals in any case.

Particular attention is drawn to the Home Office 'Employer's guide' as a mine of information. The use of a simple code can be convenient to describe the extent of compliance, the condition or the degree of control. The following Table gives an easily-understood illustration.

Code		Definition
A	Fully satisfactory	Meets all requirements, does not require improvement
B	Adequate	Some improvement possible
C	Less than adequate	Significant improvement required, action must be taken
D	Poor	Action is required urgently to improve the condition
N/A	Not applicable	Condition or circumstances not present or applicable

Alternatively the same numerical scoring system can be used as described elsewhere.

Legal summary

Major provisions are currently contained in the *Fire Precautions Act 1971*, as amended by the *Fire Safety and Safety of Places of Sport Act 1987*, and in detailed provisions concerning fire certification. The local fire authority enforces both the Fire Regulations and the parts of the *Management of Health and Safety at Work Regulations 1999 (SI 1999 No 3242)*, which deal with precautions for the safety of people where fires may occur. The latter Regulations apply to the taking of fire precautions in general terms, requiring employers to identify circumstances where situations presenting serious and imminent danger to employees could occur. These must be identified in risk assessments, and procedures written down to control risks. The *Fire Precautions (Workplace) Regulations 1997 (SI 1997 No 1840)* as amended, apply to the taking of fire precautions in premises not required to have a fire certificate.

It is anticipated that the present legal framework will change shortly with a view to simplification, but it is not considered likely that the requirements for fire risk assessment will be altered significantly.

Risk assessment – the action plan

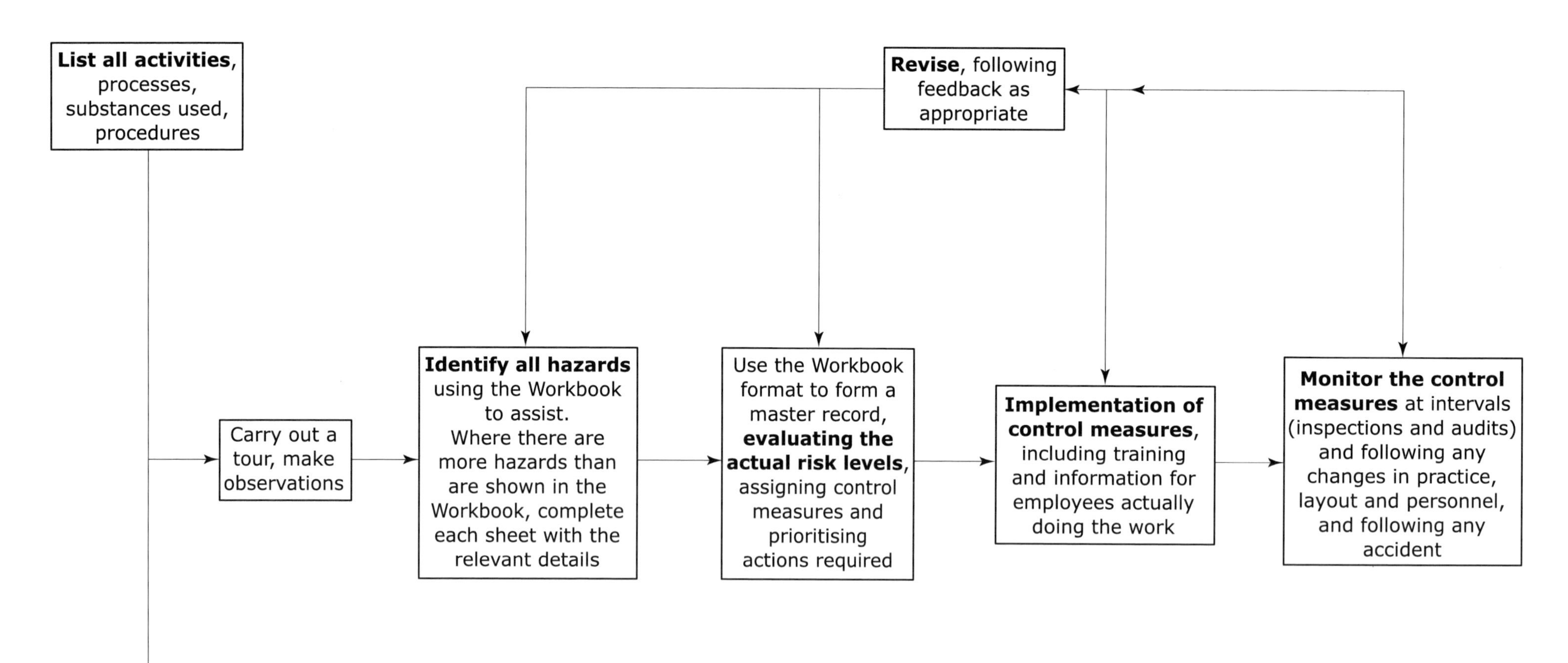

Risk Assessment in the Office

Specific requirements

The description of the office as being a high safety-risk working environment is not generally recognised as such by safety professionals, nor for that matter, by the majority of enforcers. However, as in all walks of life, legislation will affect the office working environment as it does for the major high-risk work activities.

There are no 'master office guidelines' for performing risk assessments in an office but general guidance utilising best practice is available and is described elsewhere within the workbook (see 'Application of the legislation – best practice' below).

Some office managers still view formal risk assessment in low-risk work environments as a hindrance to 'normal working'. This snap judgement is based on ignorance of office risk assessments as a management tool and a lack of understanding of what is required.

Relevant legislation

As one would expect, there are numerous health and safety Regulations which directly affect office work – each taking a different aspect and approach.

The following checklist will assist those trying to identify the legislation that applies to any office risk assessment task that has to be performed:

- *Asbestos (Licensing) Regulations 1983 (SI 1983 No 1649)* as amended: part revoked by the *Control of Asbestos at Work Regulations 2002 (SI 2002 No 2675)*
- *Confined Spaces Regulations 1997 (SI 1997 No 1713)*
- *Control of Asbestos at Work Regulations 2002 (SI 2002 No 2675)*
- *Control of Lead at Work Regulations 1998 (SI 1998 No 543)*
- *Control of Substances Hazardous to Health Regulations 2002 (SI 2002 No 2677)*
- *Electrical Equipment (Safety) Regulations1994 (SI 1994 No 3260)*
- *Electricity at Work Regulations 1989 (SI 1989 No 635)*
- *Employers' Liability (Compulsory Insurance) Regulations 1998 (SI 1998 No 2573)*
- *Employers' Health and Safety Policy Statements (Exception) Regulations 1975 (SI 1975 No 1584)*
- *Fire Certificates (Special Premises) Regulation 1976 (SI 1976 No 2003)*
- *Fire Precautions (Application for Certificate) Regulations 1989 (SI 1989 No 77)*
- *Fire Precaution (Factories, Offices, Shops and Railway Premises) Order 1989 (SI 1989 No 76)*
- *Fire Precautions (Workplace) (Amendment) Regulations 1999 (SI 1999 No 1877)*
- *Gas Safety (Installation and Use) Regulations 1998 (SI 1998 No 2451)*
- *Health and Safety (Consultation with Employees) Regulations 1996 (SI 1996 No 1513)*
- *Health and Safety (Display Screen Equipment) Regulations 1992 (SI 1992 No 2792)*
- *Health and Safety (Enforcing Authority) Regulations 1989 (SI 1989 No 1903)*
- *Health and Safety (First-Aid) Regulations 1981 (SI 1981 No 1671)* as amended by the *Health and Safety (Miscellaneous Amendments) Regulations 2002 (SI 2002 No 2174)*

- *Health and Safety Information for Employees Regulations 1989 (SI 1989 No 682)* as modified by *Health and Safety Information for Employees (Modifications and Repeals) Regulations 1995 (SI 1995 No 2923)*
- *Health and Safety (Safety Signs and Signals) Regulations 1996 (SI 1996 No 341)*
- *Health and Safety (Training for Employment) Regulations 1990 (SI 1990 No 1380)*
- *Lifting Operations and Lifting Equipment Regulations 1998 (SI 1998 No 2307)* as amended by the *Health and Safety (Miscellaneous Amendments) Regulations 2002 (SI 2002 No 2174)*
- *Manual Handling Operations Regulations 1992 (SI 1992 No 2793)* as amended by the *Health and Safety (Miscellaneous Amendments) Regulations 2002 (SI 2002 No 2174)*
- *Noise at Work Regulations1989 (SI 1989 No 1790)* modified by the *Personal Protective Equipment at Work Regulations 1992 (SI 1992 No 966)* which was again modified by the *Health and Safety (Safety Signs and Signals) Regulations 1996 (SI 1996 No 341)*
- *Notification of Cooling Towers and Evaporative Condensers Regulations 1992 (SI 1992 No 2225)*
- *Personal Protective Equipment at Work Regulations 2002 (SI 2002 No 1144)* amended by the *Health and Safety (Miscellaneous Amendments) Regulations 2002 (SI 2002 No 2174)*
- *Provision and Use of Work Equipment Regulations 1992* amended by the *Health and Safety (Miscellaneous Amendments) Regulations 2002 (SI 2002 No 2174)*
- *Reporting of Injuries, Diseases and Dangerous Occurrences Regulations 1995 (SI 1995 No 3163)*
- *Safety Representatives and Safety Committees Regulations 1977 (SI 1977 No 500)* modified by the *Management of Safety at Work Regulations 1992 (SI 1992 No 2051)*, which was again amended by the *Health and Safety (Consultation with Employees) Regulations 1996 (SI 1996 No 1513)*
- *Safety Representatives and Safety Committees Regulations 1977 (SI 1977 No 500)*
- *Social Security (Industrial Injuries) (Prescribed Diseases) Regulations 1985 (SI 1985 No 967)*
- *Workplace (Health Safety and Welfare) Regulations 1992 (SI 1992 No 3004)* as amended by the *Health and Safety (Miscellaneous Amendments) Regulations 2002 (SI 2002 No 2174)*
- *Working Time Regulations 1998 (SI 1998 No 1833)*
- *Working Time Regulations 1999 (SI 1999 No 3372) as amended by the Working Time (Amendment) Regulations 2001 (SI 2001 No 3256)*

Office assessments as a management tool

Undertaking office risk assessments will help management in the following two main areas:

- Legal compliance – surely a must?
- Cost control/reduction – an imperative to enable corporate survival!

Identifying office hazards and putting in control measures will help to:

- improve the office working methods (a two-edged benefit);
- improve output (no complaints?);
- reduce sickness/absenteeism;
- increase profit (in the long term, if not the short term);
- budget and 'cost-in' improvements over a given time period based, in part, on safety priorities.

Office risk assessment, involving participation by everyone, provides an opportunity for all concerned to agree that health and safety procedures:

- are based on shared perceptions of hazards and risks;
- are necessary and workable; and
- will succeed in preventing accidents and occupational ill-health.

Office workers' health and assessment

Asbestos

Occupational ill-health is tackled head on by risk assessment in the office; for instance, take asbestos as an example.

Asbestos in the office is no different than that within the heavy industrial or construction environment. In fact, it can pose a far greater problem in that exposure to asbestos dust can occur long before its presence in the office is detected by alterations, maintenance or formal assessment via sampling.

The Health & Safety Executive (HSE) has stated that asbestos was used extensively as a building material in Great Britain from the 1950s through to the mid 1980s. Although some of this material has been removed over the years, there are many thousands of tonnes of asbestos still present in buildings. It is estimated that over half a million non-domestic premises currently have some form of asbestos in them. Extensive repair and removal work is being undertaken which will continue for the foreseeable future.

There are several different types of asbestos, split into two groups, which can be found in offices. They are different in their chemistry and in the way that they formed millions of years ago.

Types of asbestos

The Amphibole group contains Crocidolite (blue asbestos), Amosite (brown asbestos), Anthophyllite and Tremolite. The Serpentine group contains Chrysotile (white asbestos).

Only Chrysotile (white), Crocidolite (blue), Amosite (brown) and Anthophyllite have been in common industrial use, and can be found in the following materials (not an exhaustive list):

- thermal insulation/sprayed insulation coatings/loose fill insulation;
- roofing/wall boards/guttering/external fittings;
- insulating board in electrical installations/switchgear;
- suspended ceiling tiles, plastic/vinyl floor tiles;
- gaskets/packing in plant rooms (eg glands);
- decorative finishes (eg artex, especially those approximately pre-1985);
- millboards (PF mills/backing boiler tubes/door panels);
- brake and clutch linings;
- contaminated ground, especially where asbestos materials may have been buried for disposal.

The requirement of the *Asbestos Regulations 2002* to carry out a survey of work and other premises and to perform a formal assessment of the risk from any asbestos material found, can now be clearly balanced and understood as being one of a safety management priority when looking at risk assessment in general.

Pregnant workers

Continuing the theme of occupational health, one other risk assessment naturally springs to mind when thinking of assessments in the office, namely, that which affects pregnant workers and nursing mothers.

Within the *Management of Health and Safety at Work Regulations 1999 (SI 1999 No 3242)*, two important groups of workers are clearly defined when it comes to performing risk assessments – those employees that are pregnant and those who have given birth within the previous six months or who are breastfeeding.

If, by their pregnancy, they expose themselves to risk, employers must ensure that a formal risk assessment takes place and, in particular, it is advised that the assessment should closely examine:

- any physical and mental pressure that could cause stress and anxiety with the possible result of the individual experiencing raised blood pressure;
- manual handling and the pressure this causes on the skeletal frame and muscles; and

- posture and movement of the expectant mother, eg standing for long periods, physical work of various kinds, seating associated with benchwork and display screen equipment.

If it is discovered that the existing control measures in place for individuals do not adequately avoid the risks to the mother, her body or the baby, then it would be expected that the employer makes some reasonable alteration to her work, via the hours worked or by other measures such as suitable alternative work.

Additional advice can be found in the HSE booklet HS(G)122: 'New and expectant mothers at work: A Guide for Employers'.

Hazardous substances

As thousands of people are also exposed to hazardous substances at work the *COSHH Regulations 2002*, require a degree of preventative action to be undertaken by employers and employees. If exposure is not prevented or properly controlled, it can cause serious illness, sometimes even death. Some examples of the effects of hazardous substances include:

- skin irritation, dermatitis or even skin cancer from frequent contact with oils;
- asthma resulting from sensitisation to isocyanates in paints or adhesives;
- being overcome by toxic fumes – injuries or death can be caused by the substances themselves or by falling after losing consciousness;
- poisoning by drinking toxic liquids from bottles thought to contain water or soft drinks –sometimes with fatal results;
- cancer – which can appear many years after first exposure to carcinogenic (cancer-causing) substances at work;
- infection from bacteria and other micro-organisms ('biological agents').

Types of hazardous substances

For the purposes of the *COSHH Regulations 2002*, substances hazardous to health are:

- substances or mixtures of substances classified as dangerous to health under the current *Chemicals (Hazard Information and Packaging for Supply) Regulations 2002 (SI 2002 No 1689)* (*CHIP*). These materials can be identified by their warning label and the supplier must provide a safety data sheet for them. Many, though not all, dangerous substances are listed in the HSE's publication, 'The Approved Supply List' (this publication also includes other important indicators, such as the risk phrases which help to identify substances that can cause cancer). Suppliers must decide if substances that are not in the 'Approved Supply List' are also dangerous; and, if so, to label them accordingly;
- substances with occupational exposure limits – these are listed in EH40:'Occupational exposure limits' which is revised annually;
- biological agents (bacteria and other micro-organisms) – if they are directly connected with the work or if exposure is incidental to it, such as in farming, sewage treatment or healthcare;
- any kind of dust in a substantial concentration; and
- any other substance which has comparable hazards to people's health, but which for technical reasons may not be specifically covered by the *CHIP Regulations*. Examples are some pesticides, medicines, cosmetics or intermediates produced in chemical processes.

Exceptions

It can, therefore, be seen that COSHH applies to virtually all substances hazardous to health except:

- asbestos and lead, which have their own Regulations;
- substances which are hazardous only because they are:
  - radioactive;
  - asphyxiants;
  - at high pressure;
  - at extreme temperatures; or
  - have explosive or flammable properties;

- biological agents if they are *not* directly connected with the work and they are outside the employer's control, such as catching a cold from an office colleague.

For the vast majority of proprietary chemicals the presence (or absence) of a warning label will indicate whether COSHH is relevant. For example, there is no warning label on ordinary household washing-up liquid which may be used in the office kitchen area, so if it is used, the employer does not have to worry about COSHH. However, there is a warning label on bleach and, therefore, COSHH does apply to it in the workplace.

Compliance with COSHH

Complying with the *COSHH Regulations 2002* in an office environment involves:

- **assessing** the risks to health arising from office work;
- **deciding** what precautions are needed. Do not carry out any work which could expose employees to hazardous substances without first considering the risks and the necessary precautions;
- **preventing or controlling exposure**;
- **ensuring** that the control measures are used and maintained properly, and that any safety procedures which have been laid down are followed;
- **monitoring exposure** of workers to hazardous substances and, carrying out the appropriate **health surveillance** where the assessment has shown that these are necessary or where COSHH lays down specific requirements; and
- **ensuring** that employees are properly **informed, trained and supervised**.

For substances bought into the office, the safety information should be checked and using existing knowledge of the work done in the offices, ensure current best practice is applied to their use and also check that any work-related health problems known to exist in office environments have been covered.

Contact trade associations and other employers of office staff who may be able to offer advice based on knowledge and experience.

Check whether the substance is mentioned in the *COSHH Regulations 2002*, or in other guidance such as EH40: 'Occupational exposure limits', or the approved list of biological agents. Also refer to Part V of HSE's publication, 'The Approved Supply List' (current edition), and any other available trade literature or documentation.

It is important to remember that when doing an assessment for COSHH, that it includes hazardous substances which are:

- not only brought into the office to be used, but also worked on or stored;
- given off as fumes, vapours or aerosols, or might leak or be spilled, during any process or work activity;
- produced at the end of any work or process, as finished products, waste or residues.

Assessing the risks

In order to assess the risks, consider the following:

- Manufacturers' advice on storage, use and disposal.
- Who might be affected (eg employees, contractors, the public) and what they are doing. Are they likely to be exposed to any hazardous substances present, and to what extent?
- Would such exposure involve substances being breathed in, swallowed or absorbed through the skin?
- What measures are currently taken to prevent or control exposure, and check on the effectiveness and use of those measures?
- Could leakage, spillage or release occur, eg through breakdown of the plant or controls, or through operator error?
- Existing good working practices in the office working environment that assist in any implementation of new requirements.

Remember activities sometimes overlooked in the office 9-to-5 working environment such as cleaning and maintenance – these tasks often give high chemical exposure times to workers not always considered or initially recognised as being part of the 'office teams'.

If an employer has employees who work away from their offices, similar questions must be asked about the risks at other workplaces (see HSE's guidance, 'COSHH and peripatetic workers').

Reach firm conclusions about the real risks to people's health from the information which has been gathered and the potential of the substance for causing harm.

Fire and the office

The *Fire Precautions (Workplace) Regulations 1997* (as amended) requires virtually all employers to carry out a fire risk assessment of their workplace in order to identify, reduce or eliminate any fire hazards that may be found.

Even if a fire certificate has been issued for the office premises the formal risk assessment must be completed. So what is required?

First identify the potential sources of ignition in the office and also the combustible materials that are present as part of the office working environment. Check the furnishings and also the structure in which the office work is carried out.

The main aim of any fire risk assessment is to reduce or, as far as is reasonably practicable, illuminate the risk of fire and loss of life as a result of a fire.

Opportunities to eliminate risk should also be taken during the assessment. Follow any recommendations coming from the formal written assessment to eliminate, substitute, avoid or transfer any fire hazards that have been identified.

All persons who use the office premises must be considered. These include staff, contractors, customers and members of the public. The means of escape, equipment for detecting and giving warning in case of fire and fire-fighting equipment must also be appropriate for the premises and the numbers of people present. A fire risk assessment must also account for individuals' ease of mobility. It is important, therefore, to assess the age, agility and health of the people who may be on the premises.

If the offices provide additional welfare and support programmes for staff to be used within the premises, ie a crèche, gym or fitness centre etc, the assessment must take these into account and clearly identify the fire control measures required.

The assessment control measures must also show that satisfactory escape routes are identified and kept available for use at all times and that suitable arrangements are made to detect and give warning of a fire – not forgetting that appropriate fire-fighting equipment is strategically and correctly located around the offices.

The Regulations also require that:

- employers ensure that employees are trained in the appropriate action to take when a fire breaks out, or if a fire is suspected;
- employees know how to use the fire-fighting equipment provided;
- adequate records are kept of all staff training; and
- records are kept of tests and maintenance of fire equipment.

Application of the legislation – best practice

All management actions taken regarding office risk assessments must encompass good management practice, namely:

- assess what needs to be done thoroughly, avoiding duplication or missing any opportunities for greater efficiencies;
- identify the hazards and the operational shortcomings;
- define the work methods to be employed and the skills required;
- implement the work with given time limits and resource controls;
- motivate the work and maintain active central monitoring.

The following office procedure chart as a guide through the basic principles and also applies best practice, including the performance of the task in a cost-effective manner.

Procedure	**Actions to be taken**
Hazard identification within the office	Walk around the offices and look at what could reasonably be expected to cause harm.
	List the main areas of the working environment, both internal and external.
	Draw up a list of the hazards, using collective brainstorming.
	Ignore trivial risks.
	Concentrate on the significant risks.
	Ask the question 'what if'– what if we had a fire, what if we had a spillage etc?
	Ask staff/representatives about the hazards etc.
Who might be harmed?	Consider all activities and processes undertaken by staff – routine and non-routine.
	Consider staff/contractors who conduct activities out of working hours – cleaners and maintenance etc.
	Think about people who may not be in the workplace all the time but could be harmed by office activities. Include: ● staff; ● visitors; ● contractors; ● other tenants; and ● members of the public.
Evaluating the risk	Even after all precautions have been taken, usually some risks remain: ● Decide whether the remaining risk is high, medium or low. ● If you identify a risk ask yourself the following questions: ○ Can the risk be removed completely? ○ Can a less risky option be tried? ○ Can access to the hazard-locking doors etc be prevented? ○ Can the work to reduce exposure to the hazard be organised? ● Has everything been done that is reasonably practicable to keep the workplace safe? ● Decide whether existing precautions are adequate or if more should be done. ● Aim to make all risks small by adding to the existing procedures or control measures.
Office hazard control measures	Hazard control measures should take into account the need: ● to eliminate the risk, eg employ a contract window cleaner, rather than asking the handy person to do the job; ● to substitute, eg old toners/printers with new; ● to enclose and/or fit a guard to the paper guillotine; ● to introduce a safe system of work, ie written procedures for risk areas such as reception/security; ● for supervision – ensure adequacy; ● to identify staff training needs; ● to inform staff, ie written guidance, signs, handbooks etc; ● for protection, eg issue personal protective clothing, where necessary
Recording your findings	Write down the significant office hazards and the measures to be taken to control these risks.
	Ensure that a proper check was made of the offices and work activities and that all persons who might be affected by the activity were involved.
	Ensure that all the obvious significant hazards have been dealt with, taking into account the number of people who could be involved.

	Ensure that the precautions are reasonable, and that the remaining risk(s) are low.
	Assessments need to be suitable and sufficient, not perfect.
	Keep the written assessment form for future reference.
	Ensure that the findings of the office assessment are bought to the attention of all concerned.
	Review present instructions/procedures as they may already list hazards and control measures – you may already have undertaken risk assessment and embodied the findings and control measures within these documents.
	For areas of work where similar or identical activities are carried out at a number of locations or on a number of occasions, then generic assessments may be more appropriate. Therefore, use as an aid to the efficient implementation and compliance with the Regulations.
Monitor and review the office assessments	Managers will be required to review and revise as necessary if: • there is reason to suspect the office assessment is no longer valid – this may be apparent through accidents, complaints or ill health; • there has been a significant change to matters to which the office assessment relates – this could be the introduction of a new or revised process or the introduction of new office equipment.
	It is good practise to review the assessments form from time to time.
	Ensure any changes or amendments made to the office assessments are bought to the attention of all concerned.

Risks and control measures

A typical office does not really exist as the characters and management styles alter and change. However, the type of work performed is broadly similar in most offices.

Typing, data input, copying, telephoning, attending meetings etc, with the type of accidents normally experienced, ie slips, trips and falls can be considered as the many norms.

We start with finding the existing control measures. The table below provides brief examples of typical office control measures that can be used within office environments to remove or reduce the risk(s). Use this list as a guide to produce assessments (the list is not exhaustive).

Risks	Control measures
Slips, trips and falls	Ensure standards of office housekeeping are maintained to a good standard.
	Ensure all trailing electrical cables are encased in a rubber sleeve.
	Ensure that all walk ways/corridors remain free of obstruction at all times.
	Ensure that adequate office and storage area lighting is available at all times
Falls from height	Ensure that suitable ladders or kick stools are available in all storage areas, or where access is required to office storage shelving.
	Ensure a good standard of stacking is maintained in the storage areas.
	Do not store items on the top of cupboards.
Falling objects	Do not overload storage shelving.
	Do not store items on the top of cupboards and cabinets.
Contact with hot surfaces	Ensure that all hot surfaces, ie kettles etc, are clearly marked with appropriate warning notices.
	Ensure that all hot pipes are clearly labelled with warning notices.
Contact with hot water	Ensure that all work tops etc remain free from obstructions at all times.
	Keep work tops tidy at all times.

Risks	Control measures
Electric shock	Ensure that all portable electrical appliances are subject to regular inspection and testing.
	Records to be maintained.
	Only authorised persons to undertake inspection and repair of electrical appliances.
DSE-RSI/Repetitive strain	All 'user' DSE workstations to be subject to assessment.
	Records to be maintained.
Manual handling	Suitable footwear to be worn when undertaking manual handling activities.
	Assess the situation before lifting.
	Lifting and carrying should be limited to the extent only when confident of doing so without risk of personal injury.
	Ask for assistance when required.
	Make full and proper use of handling aids, eg trolleys etc.
Contact with hazardous substances	All hazardous substances to be locked away when not in use.

Risks	Control measures
	Ensure that COSHH assessments have been undertaken and data sheets are available.
	Ensure personal protective clothing is available and worn.
Contact with moving vehicles	Ensure local speed control system is in place, ie speed ramps etc.
	Ensure traffic and pedestrian routes are clearly defined.
	All reversing company vehicles to be fitted with 'reversing warning' device.
Verbal abuse/Violence/Security	Ensure that reception areas have a panic alarm installed.
	Ensure alarm systems are subject to regular tests and service.
	Ensure that local management arrangements are in place to respond immediately to any call.
	Ensure that adequate lighting is available.
	Ensure that appropriate door control and entry security systems are in place.

Risk assessment workbook

Worked example

Toners –Various **TOLLEY'S RISK ASSESSMENT WORKBOOKS – OFFICES**

Topic	Hazard	Risk	People at Risk	Risk Rating	Control Measures	Actioned X	Residual Risk
Toners –various	Handling	Health deterioration	Staff, store keepers and	3	• **Maximum handling by trained operatives of 30 minutes per day maximum**	X	Low (rated 1)
Use/fitting	Body contami-nation	Fire and health	cleaners	3	• **Use of personal protective clothing supervised and monitored**	X	Low (rated 1)
Disposal and first aid				3	• **Toner spillage containment kits available for use – wet wipes, gloves (disposable), special waste bags**	X	Low (rated 2)
					• **Low flammability risk, however, no smoking permitted and other sources of ignition, eg heaters, sources of static electricity excluded**	X	Low (rated 2)
					• **Disposal of toners via suppliers and not treated as general office waste**	X	Low (rated 3)
					• **First-aid kits and nominated first-aiders available at all times. After significant exposure, first-aiders to call for assistance immediately. INGESTION – DO NOT INDUCE VOMITING**	X	Low (rated 1)
							Insignificant (rated 1)

Owner: AN Other

Assessment date: 10/07/2003
Re-assessment date: July 2004

Worked example (contd)

TOLLEY'S RISK ASSESSMENT WORKBOOKS – OFFICES

Questions the Risk Assessor needs to ask

No	Question	Yes	No	N/A
1	Can the job of replacing the toners be done differently?		X	
2	Are all staff used to change the toners trained and aware of the health risks?		X	
3	Are toner spillage kits available for use in a central point or on each floor and are staff aware?	X		
4	Are the toners kept secure in a locked cabinet or area prior to use?	X		
5	Is the area of the machine kept clear during toner change and/or maintenance?	X		
6				
7				
8				
9				
10				

Owner: AN Other

Assessment date: 10/07/2003
Re-assessment date: July 2004

Worked example (contd)

TOLLEY'S RISK ASSESSMENT WORKBOOKS – OFFICES

Action Plan

SECTION:

SITE:

RISK ASSESSMENT ITEM NO	ACTION REQUIRED	ACTION BY	PRIORITY 1-5	TARGET DATE	COST	COMPLETED (Name/Signature)
2	Ensure PPE stock maintained and checked regularly	AN Other	3	28/07/03	Negligible	
4	Train office cleaners in risks and safe working procedures in the event of spillage	AN Other	3	18/08/03	Approx £300 total	

KEY **Priority** **1 Urgent** **2 High** **3 Medium** **4 Low** **5 Very low**	**Action Plan prepared by (name): AN Other** **Signature:** **Date: 17/07/03**

Part 2

Pedestrians

Topic	Hazard	Risk	People at Risk	Risk Rating	Control Measures	Actioned X	Residual Risk
Pedestrian access/ egress	Personal injury	Slips, trips and falls	Employees and visitors	Low	• **Access and egress routes (footpaths and driveways) to the premises are to be maintained to a good standard, eg free from potholes and general trip hazards at all times** • **Adequate standards of lighting (artificial) should be provided, particularly to steps, stairways and passageways, during hours of darkness** • **Issues requiring attention are to be reported to identify persons (eg landlord, building manager, maintenance contractors) so that remedial action can be taken** • **Moss and lichen should be removed from walkways/paved areas on a regular basis in order to help prevent slips, trips and falls** • **Ensure that all paved areas and footpaths are safe for the use of the blind and partially sighted and wheelchair users and persons using walking aids (where applicable)**	☐ ☐ ☐ ☐ ☐ ☐ ☐ ☐ ☐ ☐	

Owner:

Assessment date:
Re-assessment date:

Questions the Risk Assessor needs to ask

No	Question	Yes	No	N/A
1	**Who set the maintenance and cleaning schedules and are they still to current needs?**	☐	☐	☐
2	**Has access standards been confirmed as adequate by enforcement and advisory agencies?**	☐	☐	☐
3	**Are cleaning materials being used?**	☐	☐	☐
4	**Do the cleaning materials require COSHH assessments and a risk assessment to be issued to cleaning employees?**	☐	☐	☐
5	**Are the lighting maintenance schedules still applicable to current needs?**	☐	☐	☐
6	**Is the lighting suitable for all conditions and security requirements?**	☐	☐	☐
7		☐	☐	☐
8		☐	☐	☐
9		☐	☐	☐
10		☐	☐	☐

Owner:

Assessment date:
Re-assessment date:

Shrubs and Trees

Topic	Hazard	Risk	People at Risk	Risk Rating	Control Measures	Actioned X	Residual Risk
Shrubs and trees	Personal injury	Slips, trips and falls	Staff and pedestrians	Low	• **Excessive plant growth to be cut back on a regular basis to prevent personal injury and/or hazard to pedestrians by obstructing or obscuring access routes**	☐ ☐ ☐ ☐ ☐ ☐ ☐ ☐ ☐ ☐	Low

Owner:

Assessment date:
Re-assessment date:

Questions the Risk Assessor needs to ask

No	Question	Yes	No	N/A
1	Are set maintenance schedules still to current needs?	☐	☐	☐
2	Do the plants and shrubs pose a significant health risk to minors if touched/eaten?	☐	☐	☐
3	What materials/equipment is being used during pruning and maintenance?	☐	☐	☐
4	Do the materials/equipment being used pose any significant risk to employees or members of the public?	☐	☐	☐
5	Have safe methods of work been formally agreed during plant and shrubbery maintenance?	☐	☐	☐
6		☐	☐	☐
7		☐	☐	☐
8		☐	☐	☐
9		☐	☐	☐
10		☐	☐	☐

Owner:

Assessment date:
Re-assessment date:

Vehicles

Topic	Hazard	Risk	People at Risk	Risk Rating	Control Measures	Actioned X	Residual Risk
Vehicle access/ egress	Contact – physical	Personal injury	Staff and visitors	Low	• **Where appropriate, speed reduction measures to be in place (sleeping policeman, speed ramps)** • **Warning notices informing drivers of site speed limit (eg 10 mph) and the location of speed reduction ramps are to be clearly displayed** • **Appropriate separation of traffic and pedestrian routes to be in place, eg footpaths, barriers, bollards, pedestrian crossings etc** • **Barriers are to be placed adjacent to pedestrian access routes such as fire exits, loading bays etc** • **Where appropriate, mirrors should be positioned around the building at 'blind' corners etc, in order to aid vehicle movement** • **Roadways to be maintained free from deep potholes etc** • **All restricted access/egress routes are to be clearly identified with appropriate signs, to warn drivers of the hazard, eg low headroom** • **All vehicle access routes to be provided with adequate lighting during the hours of darkness**	☐ ☐ ☐ ☐ ☐ ☐ ☐ ☐ ☐ ☐	Low

Owner:

Assessment date:
Re-assessment date:

Questions the Risk Assessor needs to ask

No	Question	Yes	No	N/A
1	Are the barriers and road markings including signage clear, visible and adequately maintained?	☐	☐	☐
2	Are the maintenance schedules for all of the areas being maintained current and to present needs?	☐	☐	☐
3	Are materials/equipment being used during maintenance?	☐	☐	☐
4	Do the materials/equipment being used pose any significant risk to employees or members of the public?	☐	☐	☐
5	Have safe methods of work been formally agreed during maintenance procedures?	☐	☐	☐
6	Are the lighting standards adequate for safe passage of drivers and pedestrians?	☐	☐	☐
7		☐	☐	☐
8		☐	☐	☐
9		☐	☐	☐
10		☐	☐	☐

Owner:

Assessment date:
Re-assessment date:

Underground Basement Vehicle Parking

Topic	Hazard	Risk	People at Risk	Risk Rating	Control Measures	Actioned X	Residual Risk
Under-ground basement vehicle parking	Contact with vehicles property, pedestrians, fire and explosion Asphyx's	Personal injury	Staff and visitors	Low	• **Notices to British Standards to be displayed, eg 'No Smoking', 'No Fuelling of Vehicles', 'Emergency Exits'** • **Ensure adequate fire-fighting measures available – fire extinguishers (CO2 or foam) or sprinklers** • **Ensure the local authorities have been contacted with regards to a Petroleum Licence if applicable** • **Area to be adequately ventilated – natural/mechanical** • **Notices displayed instructing drivers not to sit with engines running** • **Ensure lighting is adequate** • **Emergency lighting fitted and maintained** • **Emergency fire exit doors must remain clear and operable** • **Ensure that adequate signage is displayed indicating the means of exit in an emergency** • **Adequate measures to be in place to restrict pedestrians from vehicle access routes/ramps, eg footpaths and stairs, barriers, mirrors, signage etc**	☐ ☐ ☐ ☐ ☐ ☐ ☐ ☐ ☐ ☐	Low

Owner:

Assessment date:
Re-assessment date:

Questions the Risk Assessor needs to ask

No	Question	Yes	No	N/A
1	Are the barriers and road markings including signage clear, visible and adequately maintained?	☐	☐	☐
2	Are the maintenance schedules for all of the areas being maintained current and to present needs?	☐	☐	☐
3	Are materials/equipment being used during maintenance?	☐	☐	☐
4	Do the materials/equipment being used pose any significant risk to employees or members of the public?	☐	☐	☐
5	Have safe methods of work been formally agreed during maintenance procedures?	☐	☐	☐
6	Are the lighting standards adequate for safe passage of drivers and pedestrians?	☐	☐	☐
7	Have the fire precautions been checked and are they to current statutory requirements?	☐	☐	☐
8	Are the security standards for all users (ie CCTV, Emergency Call Points) maintained and sufficient?	☐	☐	☐
9		☐	☐	☐
10		☐	☐	☐

Owner:

Assessment date:
Re-assessment date:

Vehicle Loading Bays

Topic	Hazard	Risk	People at Risk	Risk Rating	Control Measures	Actioned X	Residual Risk
Vehicle loading bays	Contact with moving vehicles (cars, bikes etc)	Personal injury	Staff and visitors	Low	• **Barriers placed around loading bays/ramps to restrict pedestrian access into vehicle areas**	☐	Low
					• **Mirrors positioned at loading bay areas to aid reversing vehicles**	☐	
					• **Loading bay provided with at least one exit from the lower level (wider loading bays to be provided with at least two exit points, one being at each end)**	☐	
						☐	
						☐	
						☐	
						☐	
						☐	
						☐	
						☐	

Owner:

Assessment date:
Re-assessment date:

Questions the Risk Assessor needs to ask

No	Question	Yes	No	N/A
1	Are the barriers and road markings including signage clear, visible and adequately maintained?	☐	☐	☐
2	Are the bay areas maintained clear and free from obstructions/materials etc?	☐	☐	☐
3	Are the lighting standards adequate for safe passage of vehicles, drivers, loaders and pedestrians?	☐	☐	☐
4	Have safe methods of work been formally agreed and written for all loading activities?	☐	☐	☐
5	Have the safe methods of work been brought to the attention of relevant loaders and drivers?	☐	☐	☐
6		☐	☐	☐
7		☐	☐	☐
8		☐	☐	☐
9		☐	☐	☐
10		☐	☐	☐

Owner:

Assessment date:
Re-assessment date:

Private Staff Vehicle Parking Areas

Topic	Hazard	Risk	People at Risk	Risk Rating	Control Measures	Actioned X	Residual Risk
Private staff vehicle parking areas	Contact with vehicles, property, pedestrians	Personal injury	Staff	Low	• **Parking bays are to be clearly marked in order that safe access between vehicles can be maintained**	☐	Low
					• **Parking outside of designated areas is to be discouraged, eg on driveways, footpaths, in front of emergency exits etc**	☐	
					• **Place no parking signs or clamp vehicles**	☐	
					• **Car parking areas are to be provided with adequate lighting during the hours of darkness**	☐	
						☐	
						☐	
						☐	
						☐	
						☐	
						☐	

Owner:

Assessment date:
Re-assessment date:

Questions the Risk Assessor needs to ask

No	Question	Yes	No	N/A
1	**Are the barriers and road markings including signage clear, visible and adequately maintained?**	☐	☐	☐
2	**Are the parking bays maintained clear and free from obstructions/materials etc?**	☐	☐	☐
3	**Are the lighting standards adequate for safe passage of vehicles, drivers and pedestrians?**	☐	☐	☐
4	**Are the security standards adopted still current and applicable to the area?**	☐	☐	☐
5		☐	☐	☐
6		☐	☐	☐
7		☐	☐	☐
8		☐	☐	☐
9		☐	☐	☐
10		☐	☐	☐

Owner:

Assessment date:
Re-assessment date:

Site Security Team – Where Provided

Topic	Hazard	Risk	People at Risk	Risk Rating	Control Measures	Actioned X	Residual Risk
Site security team – where provided	Personal	Physical harm	Security team	Med/low	• **All external areas to be patrolled and doors and windows to be checked to ensure that they remain locked** • **Car park areas patrolled, particularly on an evening, to check for intruders** • **Serviceable torches to be available and carried at all times when undertaking security checks, especially to external areas** • **All intruder alarms (if fitted) are to be maintained in good working order at all times** • **Public Address (PA) system (where fitted) to be routinely checked to ensure that it is working at all times** • **Police to be contacted immediately if an intruder is on the grounds or in the building** • **Report all incidents to the line manager/supervisor and site security** • **Door entry security (where fitted) codes to be changed on a regular basis** • **Security swipe passes (where used) to be carried at all times** • **Procedures are to be in place for informing security personnel of employees who are working after hours and/or lone working**	☐ ☐ ☐ ☐ ☐ ☐ ☐ ☐ ☐ ☐	Low

Owner:

Assessment date:
Re-assessment date:

Questions the Risk Assessor needs to ask

No	Question	Yes	No	N/A
1	Are the security teams adequately trained and medically fit for security activities?	☐	☐	☐
2	Are the security standards for the premises and the team controlled, monitored and regularly assessed?	☐	☐	☐
3	Are the lighting standards for all areas to good standards and maintained?	☐	☐	☐
4	Are the security teams tested as to their knowledge of the security standards?	☐	☐	☐
5	Have the PPE for the teams been selected in accordance with known risk and personal requirements?	☐	☐	☐
6	If the security team is a contracted-out service, is the contractor a member of the relevant trade association?	☐	☐	☐
7	Has the contractor been adequately vetted?	☐	☐	☐
8	Do the teams have current contact lists and emergency numbers?	☐	☐	☐
9	Is the communications equipment sufficient and does it cover all of the premises (no 'black spots')?	☐	☐	☐
10	Are the security logs maintained current and are they applicable to current needs?	☐	☐	☐

Owner:

Assessment date:
Re-assessment date:

Entrance Doors

Topic	Hazard	Risk	People at Risk	Risk Rating	Control Measures	Actioned X	Residual Risk
Entrance doors	Contact injury	Personal injury/bruising etc	Staff and visitors	Low	• **All doors to be adequately maintained to ensure safe operation, eg free movement, no sharp edges on handles doors or frame**	☐	Low
					• **Glass doors, subject to risk assessment, to have safety glass fitted whereby danger may be evident should a person come into contact, eg swing doors etc.**	☐	
					• **Glass doors to have large motif, etching or transfer to help reduce the risk of persons walking into them**	☐	
					• **Revolving doors to have rubber edging strips to prevent fingers becoming crushed/trapped between door frame**	☐	
					• **Automatic revolving doors to be regularly maintained and checked to ensure that motor sensor devices are working correctly**	☐	
					• **Automatic revolving doors to be fitted with an override device in the event of fire or an alternative escape door provided adjacent to the revolving doors**	☐	
						☐	
						☐	
						☐	
						☐	

Owner:

Assessment date:
Re-assessment date:

Questions the Risk Assessor needs to ask

No	Question	Yes	No	N/A
1	Are the doors maintained to current and suitable standards?	☐	☐	☐
2	Is the glass fitted to current standards and/or in sound condition?	☐	☐	☐
3	Are door maintenance schedules maintained in accordance with need?	☐	☐	☐
4		☐	☐	☐
5		☐	☐	☐
6		☐	☐	☐
7		☐	☐	☐
8		☐	☐	☐
9		☐	☐	☐
10		☐	☐	☐

Owner:

Assessment date:
Re-assessment date:

Reception Desk

Topic	Hazard	Risk	People at Risk	Risk Rating	Control Measures	Actioned X	Residual Risk
Reception desk	Individual security	Personal injury and attack	Staff and visitors	Low	• **All visitors to be signed into the visitors book/security passes issued**	☐	Low
					• **Visitors to be notified to relevant members of staff and accompanied whilst on the premises**	☐	
					• **Visitors to be briefed on emergency arrangements or details made available on reverse of security pass etc**	☐	
					• **Adequate arrangements based on risk assessment, to be established for the safety of staff, eg panic alarms, security screens, CCTV**	☐	
					• **Reception staff to be briefed on action to take in event of emergency situations, eg fire, bomb threat etc**	☐	
					• **Notice to be displayed with full details of emergency contact numbers, premise address (including postcode)**	☐	
						☐	
						☐	
						☐	
						☐	

Owner:

Assessment date:
Re-assessment date:

Questions the Risk Assessor needs to ask

No	Question	Yes	No	N/A
1	Are items that are delivered and temporarily stored in reception kept in a safe and secure area and are the staff aware of the contents and recipient?	☐	☐	☐
2	Have all the relevant reception staff – full- and part-time, been trained in the emergency procedures?	☐	☐	☐
3	Are the emergency procedures, including security procedures, checked regularly?	☐	☐	☐
4	Have all the relevant staff easy access to emergency numbers and staff contacts?	☐	☐	☐
5	Have reception staff been trained in how to avoid violence and aggression?	☐	☐	☐
6	Are the signing-in procedures for visitors maintained and safety information given to visitors?	☐	☐	☐
7	Do the reception staff know the whereabouts of the trained first-aiders?	☐	☐	☐
8	Are the reception housekeeping and cleaning standards maintained at a high level?	☐	☐	☐
9	Are all tests and checks logged?	☐	☐	☐
10	Is the risk assessment still valid	☐	☐	☐

Owner:

Assessment date:
Re-assessment date:

Topic	Hazard	Risk	People at Risk	Risk Rating	Control Measures	Actioned X	Residual Risk
Material Storage (including substances hazardous to health)	Personal contact, fire and manual lifting, personal health	Personal injury and health damage	Staff	Low	• **Boxes, paper, and other material not to be stored on top of lockers, cabinets or filing cabinets**	☐	Low
					• **Boxes and cartons to be of uniform size in any pile or stack**	☐	
					• **Heavy objects to be stored at low level to mitigate awkward lifting**	☐	
					• **Do not place objects or stacked material on windowsills**	☐	
					• **Materials to be stored inside cabinets, files, and lockers where practicable**	☐	
					• **Office equipment such as typewriters, index files, files calculators etc, be kept away from edge of desks, cabinets or tables**	☐	
					• **Aisles, corners and passageways not to be used as stacking or storage points even for temporary purposes**	☐	
					• **Designated storage areas only to be used for that purpose and areas to be locked when not in use**	☐	
					• **Safe handling and use information (date sheets) to be kept in the storage facility for inspection and use by staff**	☐	
					• **Flammable or substances hazardous to health should not be stored near heat sources or in kitchen/food preparation areas**	☐	

Owner:

Assessment date:
Re-assessment date:

Questions the Risk Assessor needs to ask

No	Question	Yes	No	N/A
1	Has suitable and safe storage areas and equipment been identified?	☐	☐	☐
2	Are staff aware of the storage methods and housekeeping standards?	☐	☐	☐
3	Are the COSHH safety date sheets and applicable risk assessments been obtained and issued or brought to the attention of relevant staff?	☐	☐	☐
4	Have any training requirements been identified for staff and are they being actioned?	☐	☐	☐
5	Is the time period for training reasonable given the known levels of risk?	☐	☐	☐
6	Are the staff aware of any written safety precautions for the work area and are they current and fully understood by all staff?	☐	☐	☐
7		☐	☐	☐
8		☐	☐	☐
9		☐	☐	☐
10		☐	☐	☐

Owner:

Assessment date:
Re-assessment date:

Topic	Hazard	Risk	People at Risk	Risk Rating	Control Measures	Actioned X	Residual Risk
Visual display units	Ergonomics and personal health	Skeletal and muscle damage	Staff	Low	• **Ensure that display screen equipment (DSE) workstations assessed in accordance with *Health and Safety (Display Screen Equipment) Regulations 1992 (SI 1992 No 2792)***	☐	Low
					• **Ensure that all DSE meets with minimum requirements**	☐	
					• **Eye tests to be afforded by the employer and sufficient monies allocated for the purchase of a basic pair of frames and lenses if required**	☐	
					• **Ensure that the employer makes staff aware that free eye tests are available**	☐	
					• **Keep all display screen equipment risk assessments on individual personal files**	☐	
					• **Keep all records of eye tests and purchase details regarding spectacles**	☐	
					• **Ensure that staff have received suitable information, instruction and training when using DSE**	☐	
						☐	
						☐	
						☐	

Owner:

Assessment date:
Re-assessment date:

Questions the Risk Assessor needs to ask

No	Question	Yes	No	N/A
1	Have the DSE assessments been completed for all staff using DSE equipment on a regular basis?	☐	☐	☐
2	Have the assessors for DSE been trained and is the training suitable and sufficient?	☐	☐	☐
3	Are records of the assessments kept up to date including actions and decisions?	☐	☐	☐
4	Are staff aware of the standards set for DSE use and the welfare (optical) arrangements?	☐	☐	☐
5	Is the furniture supplied including the chairs provided for staff using DSE suitable and sufficient?	☐	☐	☐
6	Have DSE assessments and standards for pregnant staff been fully understood by line management?	☐	☐	☐
7		☐	☐	☐
8		☐	☐	☐
9		☐	☐	☐
10		☐	☐	☐

Owner:

Assessment date:
Re-assessment date:

Topic	Hazard	Risk	People at Risk	Risk Rating	Control Measures	Actioned X	Residual Risk
Office furniture and equipment	Slips, trips and falls	Falls and physical damage	Staff	Low	• **Office furniture, equipment and electrical appliances to be arranged in order to obtain maximum safety and use of the facilities** • **Desks, file cabinets etc, to be arranged so that drawers do not open into aisles or walkways** • **File cabinet drawers to be closed after each use and only one drawer at a time to be opened to prevent toppling** • **The weight of documents etc, in the file cabinets to be evenly distributed** • **Free standing cabinets, bookcases, and shelving units to be secured to the walls/floors to prevent movement** • **Faulty desks, chairs, or other office equipment must be taken out of service or repaired immediately** • **Knife blades (guillotines) must have guards fitted** • **Photocopiers to be located in well-ventilated areas** • **Proper and regular maintenance must be provided for all mechanical and electrical equipment** • **Barriers, partitions or dividers to be properly anchored to the floor and/or made stable and positioned to afford the best noise reduction from staff and equipment**	☐ ☐ ☐ ☐ ☐ ☐ ☐ ☐ ☐ ☐	Low

Owner:

Assessment date:
Re-assessment date:

Questions the Risk Assessor needs to ask

No	Question	Yes	No	N/A
1	Is the office furniture in a good and sound condition?	☐	☐	☐
2	Are regular checks of the furniture made by competent staff or management?	☐	☐	☐
3	Is defective furniture immediately removed from use and, in particular, chairs and shelving systems?	☐	☐	☐
4	Is the office equipment regularly checked to ensure safe operation?	☐	☐	☐
5	Is defective equipment immediately removed from the office/work area and suitably marked to show it is defective and not to be used?	☐	☐	☐
6	Are staff aware of the maintenance and inspection procedures including safe operation of the equipment?	☐	☐	☐
7	Are training requirements for staff identified and actioned?	☐	☐	☐
8		☐	☐	☐
9		☐	☐	☐
10		☐	☐	☐

Owner:

Assessment date:
Re-assessment date:

Topic	Hazard	Risk	People at Risk	Risk Rating	Control Measures	Actioned X	Residual Risk
Toilets and washing facilities	Slips, trips and falls	Personal injury, health damage	Staff and visitors	Low	• **Accessible, suitable and sufficient clean toilet facilities must be provided**	☐	Low
					• **Ideally, separate toilet facilities should be provided for male and females**	☐	
					• **Accessible, clean washing facilities to be available within the welfare facility**	☐	
					• **Soap and suitable drying facilities to be provided at all times**	☐	
					• **Suitable sanitary dispensers to be made available within the female toilets and emptied in accordance with the waste regulations currently in force**	☐	
						☐	
						☐	
						☐	
						☐	
						☐	

Owner:

Assessment date:
Re-assessment date:

Questions the Risk Assessor needs to ask

No	Question	Yes	No	N/A
1	Are the toilets and associated equipment and facilities regularly cleaned?	☐	☐	☐
2	Is the housekeeping standards maintained at a high level including the air extraction equipment?	☐	☐	☐
3	Are the standards monitored and cleaning schedules maintained in accordance with the needs of the organisation?	☐	☐	☐
4	Are safe working methods employed by the cleaners/contractors?	☐	☐	☐
5	Are the cleaning materials and equipment kept in a safe and secure manner?	☐	☐	☐
6	Are COSHH data sheets and risk assessments available for relevant staff needs and are they understood?	☐	☐	☐
7	Are showers (if provided) regularly cleaned, including shower heads, to relevant safe standards?	☐	☐	☐
8		☐	☐	☐
9		☐	☐	☐
10		☐	☐	☐

Owner:

Assessment date:
Re-assessment date:

Topic	Hazard	Risk	People at Risk	Risk Rating	Control Measures	Actioned X	Residual Risk
Office – general flooring	Slips, trips and falls	Personal injury	Staff	Low	• **Regular flooring checks to be made by cleaners or supervisors** • **Any flooring defects, ie split carpets, tiles or lifting floor covering to be immediately rectified if in the main access and egress routes** • **Temporary repairs to be instigated only as an interim measure, eg tape** • **Information regarding defects or repairs to be passed to staff and, if practicable, safety representatives** • **Trailing leads and cables across main walk areas to be prevented or protected in such a manner as to prevent any slips, trips or falls**	☐ ☐ ☐ ☐ ☐ ☐ ☐ ☐ ☐ ☐	Low

Owner:

Assessment date:
Re-assessment date:

Questions the Risk Assessor needs to ask

No	Question	Yes	No	N/A
1	Have flooring maintenance and cleaning risk assessments been undertaken?	☐	☐	☐
2	If utilised, are the maintenance and repairs contractors competent?	☐	☐	☐
3	Have formal inspections by the organisation been actioned?	☐	☐	☐
4	Is documentation available for inspection regarding repairs and maintenance?	☐	☐	☐
5	Have any complaints been made or accidents been reported involving unsafe flooring or condition?	☐	☐	☐
6		☐	☐	☐
7		☐	☐	☐
8		☐	☐	☐
9		☐	☐	☐
10		☐	☐	☐

Owner:

Assessment date:
Re-assessment date:

Topic	Hazard	Risk	People at Risk	Risk Rating	Control Measures	Actioned X	Residual Risk
Lighting	Slips, trips and falls, personal health, fire, electrocution	Personal injury	Staff and visitors	Low	• **A preventative maintenance programme for all office lighting is to be established** • **All defective light bulbs or florescent tubes to be repaired immediately upon notification of defects** • **Regular maintenance checks to be recorded** • **Glare within the office areas is to be reduced so far as is reasonably practicable** • **Only trained and authorised persons deemed competent to carry out repairs to defective lighting equipment of any description** • **A full IEE inspection and test certificate must be available for the complete electrical installation. This must be renewed every five years**	☐ ☐ ☐ ☐ ☐ ☐ ☐ ☐ ☐ ☐	Low

Owner:

Assessment date:
Re-assessment date:

Questions the Risk Assessor needs to ask

No	Question	Yes	No	N/A
1	Are electrical contractors competent and accredited?	☐	☐	☐
2	Is the maintenance schedule in accordance with current needs?	☐	☐	☐
3	If applicable have permit-to-work systems and procedures been agreed prior to commencement of any electrical work?	☐	☐	☐
4	Have written safe systems of work been submitted and agreed with the organisation prior to commencement of work?	☐	☐	☐
5	Have risk assessments been undertaken and at risk staff been identified and informed of control measures?	☐	☐	☐
6		☐	☐	☐
7		☐	☐	☐
8		☐	☐	☐
9		☐	☐	☐
10		☐	☐	☐

Owner:

Assessment date:
Re-assessment date:

Topic	Hazard	Risk	People at Risk	Risk Rating	Control Measures	Actioned X	Residual Risk
Emergency planning and first aid	Personal safety, corporate safety	Personal injury	Staff and visitors	Low	• **Procedures to be adopted for any injury, illness, fire or other emergencies that affect the office staff and the working environment**	☐	Low
					• **Ensure that first aid facilities provided (including trained first aiders) match the known risks to staff whilst working**	☐	
					• **Procedures to be in place for reporting and recording all accidents in the workplace**	☐	
					• **Ensure that the emergency services can gain easy access to all parts of the office and workplace**	☐	
					• **Ensure that emergency evacuation procedures are in place and rehearsed**	☐	
					• **Assign responsibilities to competent staff who can assist in the event of any emergency incident**	☐	
					• **Ensure that staff nominated to assist are fully trained in the required duties that they are expected to perform**	☐	
						☐	
						☐	
						☐	

Owner:

Assessment date:
Re-assessment date:

Questions the Risk Assessor needs to ask

No	Question	Yes	No	N/A
1	Have trained first aiders been appointed?	☐	☐	☐
2	Are staff aware of the names and work areas of the first aiders?	☐	☐	☐
3	Have appointed persons been agreed and confirmed?	☐	☐	☐
4	Are the first-aid facilities provided allied to the current work risks that staff are exposed to?	☐	☐	☐
5	Are the facilities provided, including any first-aid room, maintained in a clean and healthy condition?	☐	☐	☐
6	Are the staff aware of the emergency procedures?	☐	☐	☐
7	Have the emergency services been informed of any special risk in the premises?	☐	☐	☐
8	Have fire and emergency evacuation marshals been appointed?	☐	☐	☐
9	Are the evacuation procedures exercised regularly and in accordance with any fire certificate?	☐	☐	☐
10		☐	☐	☐

Owner:

Assessment date:
Re-assessment date:

Topic	Hazard	Risk	People at Risk	Risk Rating	Control Measures	Actioned X	Residual Risk
Contractors	Personal and general	Injury		Med/low	• **All contractors to be signed-in as required for visitors**	☐	Low
					• **Contractors to be selected only following checks on ability (references) and safety compliance procedures**	☐	
					• **A signed works order or prior notice of attendance to be available for checking if requested by the reception or other nominated staff**	☐	
					• **The office/premises contact to be informed of attendance and instructions obtained before commencement of any work or admittance gained**	☐	
					• **All safety and relevant security information to be given to contractors in writing**	☐	
						☐	
						☐	
						☐	
						☐	
						☐	

Owner:

Assessment date:
Re-assessment date:

Questions the Risk Assessor needs to ask

No	Question	Yes	No	N/A
1	Are contractors used by the organisation accredited and competent?	☐	☐	☐
2	Are procedures for contractor vetting current and actively monitored?	☐	☐	☐
3	Have contractors been informed of the security and safety standards that they must adhere to prior to selection and any work activity?	☐	☐	☐
4	Have safety procedures and risk assessments been actively checked prior to the commencement of any work by contractors?	☐	☐	☐
5		☐	☐	☐
6		☐	☐	☐
7		☐	☐	☐
8		☐	☐	☐
9		☐	☐	☐
10		☐	☐	☐

Owner:

Assessment date:
Re-assessment date:

Topic	Hazard	Risk	People at Risk	Risk Rating	Control Measures	Actioned X	Residual Risk
Manual handling	Trips, and falls	Personal injury and health damage	Staff	Med/low	• **Staff engaged on manual lifting tasks to be trained to do so in a safe and practical manner** • **Training to include lifting assessments, practical exercises and demonstrations** • **Details of training and staff attending to be kept on personal files for inspection and reference** • **Staff should ensure that they obtain help or use trolleys when any large or awkward/heavy object has to be moved** • **When lifting and carrying loads, staff must make sure that their path is clear and that there is a safe place to put down the load**	☐ ☐ ☐ ☐ ☐ ☐ ☐ ☐ ☐ ☐	Low

Owner:

Assessment date:
Re-assessment date:

Questions the Risk Assessor needs to ask

No	Question	Yes	No	N/A
1	Have staff who are used for manual lifting activities been adequately trained?	☐	☐	☐
2	Are those staff medically and physically fit to perform the task?	☐	☐	☐
3	Is the training updated and applicable to the current needs of the organisation?	☐	☐	☐
4	Are the trainers accredited and suitable?	☐	☐	☐
5	Have the manual lifting risk assessments been undertaken and the information regarding risks and control measures been passed to all relevant staff?	☐	☐	☐
6	Has lifting equipment been purchased to assist heavy lifting tasks?	☐	☐	☐
7	Have the relevant staff been trained to use the equipment in a safe manner?	☐	☐	☐
8		☐	☐	☐
9		☐	☐	☐
10		☐	☐	☐

Owner:

Assessment date:
Re-assessment date:

Maintenance

Topic	Hazard	Risk	People at Risk	Risk Rating	Control Measures	Actioned X	Residual Risk
Lift cars – mainte-nance	Isolation Slips and falls	Lone working Falls from height	Staff and visitors	Low	• **Regular lift maintenance is to be undertaken by contractors, under a service agreement** • **Appropriate safety warnings signs (ie 'lift out of service') to be provided by the lift engineers and displayed on all landings/in the machine room when working on the lift** • **Lift engineer to provide and erect suitable and sufficient barriers whenever it is necessary to unlock or open a lift-landing door** • **Landing doors must not be left open any longer than is necessary and not left open whilst unattended** • **Records of all lift inspections, maintenance and statutory test and thorough examinations are to be maintained by the premise manager and/or landlord**	☐ ☐ ☐ ☐ ☐ ☐ ☐ ☐ ☐ ☐	Low

Owner:

Assessment date:
Re-assessment date:

Questions the Risk Assessor needs to ask

No	Question	Yes	No	N/A
1	Are the lift engineers used for maintenance accredited and recognised as competent?	☐	☐	☐
2	Have risk assessments been supplied by the lift engineers?	☐	☐	☐
3	Have safe working methods been agreed prior to commencement of any work?	☐	☐	☐
4	Is the documentation regarding defects, repairs and services readily available for inspection?	☐	☐	☐
5	Are the reception staff informed of maintenance activities or defects to lifts?	☐	☐	☐
6	Are staff aware of the procedure to report defects to lifts?	☐	☐	☐
7	Are the lift rooms maintained locked at all times and notices informing staff that the rooms are access restricted fixed to the doors?	☐	☐	☐
8		☐	☐	☐
9		☐	☐	☐
10		☐	☐	☐

Owner:

Assessment date:
Re-assessment date:

Release of Trapped Person(s)

Topic	Hazard	Risk	People at Risk	Risk Rating	Control Measures	Actioned X	Residual Risk
Lift cars – release of trapped person(s)	Manual handling	Skeletal and muscle damage	Staff	Low	• **Premise managers or staff are not to undertake the release of trapped person(s), unless trained to do so** • **The emergency services or lift engineer to be called out where staff are not adequately trained to deal with emergencies**	☐ ☐ ☐ ☐ ☐ ☐ ☐ ☐ ☐ ☐	Low

Owner:

Assessment date:
Re-assessment date:

Questions the Risk Assessor needs to ask

No	Question	Yes	No	N/A
1	Do staff know the reporting procedures regarding trapped persons in lift cars?	☐	☐	☐
2	Are the emergency call numbers fixed in the lift cars and are the emergency call facilities regularly checked?	☐	☐	☐
3	Are the contractors used for the release of persons available for immediate response during all work hours?	☐	☐	☐
4		☐	☐	☐
5		☐	☐	☐
6		☐	☐	☐
7		☐	☐	☐
8		☐	☐	☐
9		☐	☐	☐
10		☐	☐	☐

Owner:

Assessment date:
Re-assessment date:

Safe Operation

Topic	Hazard	Risk	People at Risk	Risk Rating	Control Measures	Actioned X	Residual Risk
Lift cars – safe operation	Slips, trips and falls	Injury	Staff and visitors	Low	• **Only trained lift engineers to repair or alter the electrical and/or mechanical equipment of the lift car**	☐	Low
					• **Warning notices – 'Do not use lift in an emergency situation' – to be displayed adjacent to the lift doors**	☐	
					• **Door closing mechanisms on lift car to be monitored to ensure that they do not close too fast or strongly which may cause injury**	☐	
					• **Door-closing sensors to be periodically tested and monitored**	☐	
					• **Floor level tolerances to be monitored to ensure that the lift car comes to rest at the required position and does not create a trip hazard**	☐	
						☐	
						☐	
						☐	
						☐	
						☐	

Owner:

Assessment date:
Re-assessment date:

Questions the Risk Assessor needs to ask

No	Question	Yes	No	N/A
1	Are the lift cars regularly checked for safe operation by the engineers?	☐	☐	☐
2	Are the emergency call facilities regularly checked and documented?	☐	☐	☐
3	Is the documentation available for immediate inspection?	☐	☐	☐
4		☐	☐	☐
5		☐	☐	☐
6		☐	☐	☐
7		☐	☐	☐
8		☐	☐	☐
9		☐	☐	☐
10		☐	☐	☐

Owner:

Assessment date:
Re-assessment date:

Topic	Hazard	Risk	People at Risk	Risk Rating	Control Measures	Actioned X	Residual Risk
Escalators	Slips , trips and falls	Personal injury	Staff and visitors	Low	• **Ensure that a maintenance contract has been established for the regular inspection, testing and maintenance of the escalators to be carried out by a competent person** • **Inspections to be carried out every six months and records maintained** • **Escalators to be fitted with emergency stop buttons which are prominent, well marked and easy to reach** • **Ensure that emergency controls are subject to regular tests and records of tests maintained** • **All safety signage to be made prominent and must be to current British Standard**	☐ ☐ ☐ ☐ ☐ ☐ ☐ ☐ ☐ ☐	Low

Owner:

Assessment date:
Re-assessment date:

Questions the Risk Assessor needs to ask

No	Question	Yes	No	N/A
1	Are accredited escalator engineers used for maintenance and repair?	☐	☐	☐
2	Are the emergency stop facilities regularly checked and documented?	☐	☐	☐
3	Are staff aware of the defect reporting procedures?	☐	☐	☐
4	Are written emergency procedures current and in accordance with requirements?	☐	☐	☐
5	Are the maintenance areas and escalator equipment maintained and clean with no risk of fire?	☐	☐	☐
6	Have risk assessments been completed if considered necessary?	☐	☐	☐
7	Have any accidents or incidents been reported which involve the escalators and what is the result of investigations?	☐	☐	☐
8		☐	☐	☐
9		☐	☐	☐
10		☐	☐	☐

Owner:

Assessment date:
Re-assessment date:

Topic	Hazard	Risk	People at Risk	Risk Rating	Control Measures	Actioned X	Residual Risk
Gas appliance servicing/ mainte-nance	Fire and explosion	Personal injury	Staff	Low	• **No person (including contractors) to carry out any work in relation to a gas fitting/installation or appliance, unless CORGI registered** • **Proof of registration to be obtained prior to work or maintenance contract placement** • **Annual gas safety checks to be carried out on all gas appliances and copies of 'Landlords Gas Safety' certificates retained on site**	☐ ☐ ☐ ☐ ☐ ☐ ☐ ☐ ☐ ☐	Low

Owner:

Assessment date:
Re-assessment date:

Questions the Risk Assessor needs to ask

No	Question	Yes	No	N/A
1	Are the contractors used for gas appliance maintenance and repair CORGI registered?	☐	☐	☐
2	Have safe working procedures been agreed prior to commencement of any work by the contractor?	☐	☐	☐
3	Are staff aware of emergency procedures regarding the reporting of defective gas appliances and equipment?	☐	☐	☐
4	Is the safety documentation and certification available for inspection?	☐	☐	☐
5		☐	☐	☐
6		☐	☐	☐
7		☐	☐	☐
8		☐	☐	☐
9		☐	☐	☐
10		☐	☐	☐

Owner:

Assessment date:
Re-assessment date:

Topic	Hazard	Risk	People at Risk	Risk Rating	Control Measures	Actioned X	Residual Risk
Gas boiler rooms	Explosion, fire Slips Slips, trips and falls Contact with hot surfaces Hazardous substances	Personal injury and health damage	Staff and contractors	Low	• **Plant rooms to be designated 'No smoking' and entry by 'authorised persons only'** • **Plant rooms should preferably be fitted with gas leakage detectors, which are to be periodically checked by competent persons, eg annually** • **Plant rooms are not to be used for general storage** • **Fire control measures such as extinguishers, smoke/heat detectors, sprinklers, automatic shut-off valves etc to be subject to regular service by contractors** • **Master gas control switches to isolate gas supplies to be clearly marked, eg painted yellow**	☐ ☐ ☐ ☐ ☐ ☐ ☐ ☐ ☐ ☐	Low

Owner:

Assessment date:
Re-assessment date:

Questions the Risk Assessor needs to ask

No	Question	Yes	No	N/A
1	Are gas intake and boiler rooms restricted as to access and maintained secure at all times?	☐	☐	☐
2	Are emergency procedures and call numbers known to all relevant staff?	☐	☐	☐
3	Are the areas maintained free from obstructions and cleaned regularly?	☐	☐	☐
4	Are schematic drawings available for emergency service use in the event of any incident?	☐	☐	☐
5	Is the lighting suitable and maintained in all areas?	☐	☐	☐
6	Are the bunded areas, if constructed, maintained clean and free from debris?	☐	☐	☐
7	Are all pipes colour coded and signed?	☐	☐	☐
8	Is the maintenance and defect documentation maintained current and available for inspection?	☐	☐	☐
9		☐	☐	☐
10		☐	☐	☐

Owner:

Assessment date:
Re-assessment date:

Topic	Hazard	Risk	People at Risk	Risk Rating	Control Measures	Actioned X	Residual Risk
Service cupboards – gas and electricity	Fire, Electric Shock	Personal Injury	Staff	Low	• **Service cupboards not to be used for general storage**	☐	Low
					• **Service cupboards to be locked when not in use**	☐	
					• **Service cupboards to display appropriate safety notices – ie gas intake, main electrical intake, fireman's switch etc**	☐	
					• **Clear access to be maintained to all service cupboards at all times**	☐	
						☐	
						☐	
						☐	
						☐	
						☐	
						☐	

Owner:

Assessment date:
Re-assessment date:

Questions the Risk Assessor needs to ask

No	Question	Yes	No	N/A
1	Are service cupboards maintained secure at all times and signed indicating restricted access?	☐	☐	☐
2	Are the service cupboards maintained clean and free from obstructions?	☐	☐	☐
3	Are access keys always and easily available/obtained in an emergency?	☐	☐	☐
4	Are relevant staff aware of the safe working procedures and emergency procedures for incidents involving the service cupboards?	☐	☐	☐
5		☐	☐	☐
6		☐	☐	☐
7		☐	☐	☐
8		☐	☐	☐
9		☐	☐	☐
10		☐	☐	☐

Owner:

Assessment date:
Re-assessment date:

Topic	Hazard	Risk	People at Risk	Risk Rating	Control Measures	Actioned X	Residual Risk
Restrooms and meal facilities	Personal	Personal health and comfort	Staff	Low	• **Ensure that restrooms provide for the protection of non-smokers from discomfort caused by tobacco smoke** • **Ensure that any areas provided for catering machines and drink dispensers are maintained free from obstruction and that rubbish containers are regularly emptied** • **Suitable facilities for pregnant workers and nursing mothers to be available** • **A clean supply of drinking water to be supplied for use within the rest or meal/drink dispensing areas**	☐ ☐ ☐ ☐ ☐ ☐ ☐ ☐ ☐ ☐	Low

Owner:

Assessment date:
Re-assessment date:

Questions the Risk Assessor needs to ask

No	Question	Yes	No	N/A
1	Are rest areas maintained clean at all times?	☐	☐	☐
2	Are the facilities for smoke extraction in smoke-free areas maintained?	☐	☐	☐
3	Are the smoke-free areas monitored to ensure compliance?	☐	☐	☐
4	Have pregnant and nursing mother areas been agreed and are they available for use?	☐	☐	☐
5	Is clean drinking water supplied at all times?	☐	☐	☐
6		☐	☐	☐
7		☐	☐	☐
8		☐	☐	☐
9		☐	☐	☐
10		☐	☐	☐

Owner:

Assessment date:
Re-assessment date:

Internal

Topic	Hazard	Risk	People at Risk	Risk Rating	Control Measures	Actioned X	Residual Risk
Internal smoking areas	Personal health	Health and personal injury	Staff	Low	• **Designated smoking areas are to be provided with:**	☐	Low
					○ heat detectors	☐	
					○ fire extinguishers	☐	
					○ flame-resistant furnishings	☐	
					○ metal/glass ashtrays	☐	
					• **Smoking areas to be well ventilated to open air and no smoke must transfer into the designated non-smoking areas**	☐	
						☐	
						☐	
						☐	
						☐	

Owner:

Assessment date:
Re-assessment date:

Questions the Risk Assessor needs to ask

No	Question	Yes	No	N/A
1	Have smoking areas been agreed?	☐	☐	☐
2	Is the smoke extraction maintained at all times and exhausted to open air?	☐	☐	☐
3	Is the area(s) maintained free from the accumulation of waste and smoking materials?	☐	☐	☐
4	Is the fire emergency systems maintained and sufficient for current needs within the smoking areas?	☐	☐	☐
5	Is the furniture supplied for the area to current British Standards?	☐	☐	☐
6		☐	☐	☐
7		☐	☐	☐
8		☐	☐	☐
9		☐	☐	☐
10		☐	☐	☐

Owner:

Assessment date:
Re-assessment date:

Topic	Hazard	Risk	People at Risk	Risk Rating	Control Measures	Actioned X	Residual Risk
Office Stress	Personal health	Long-term absence and health affected	Staff	Low/ med	• **All staff are to be provided with relevant information and workloads assessed in order to mitigate the effects of stress. Where reasonable and practicable:**	☐	Low
					○ reduce irritant noise levels	☐	
					○ be flexible in work tasks and discuss changes that affect individuals to lessen anxiousness	☐	
					○ introduce controls over work processes and the pace of operations	☐	
					○ mitigate if possible the feeling of job insecurity	☐	
					○ support staff and colleagues who deal with difficult or abusive clients or colleagues	☐	
					○ delegate work where practicable	☐	
					○ staff should learn to discuss and set limits of work input with the supervisors/managers	☐	
					○ balance work levels where practicable	☐	
					○ limit where practicable boring and monotonous tasks	☐	

Owner:

Assessment date:
Re-assessment date:

Questions the Risk Assessor needs to ask

No	Question	Yes	No	N/A
1	Is stress considered to be a particular problem to the organisation?	☐	☐	☐
2	Is work allocation to staff monitored to ensure reasonable work loads?	☐	☐	☐
3	Is stress counselling for all staff provided if required by the organisation?	☐	☐	☐
4	Are absence rates monitored and assessed for signs of stress related staff absence?	☐	☐	☐
5	Are relevant staff trained to recognise-stress related disorders or behaviour?	☐	☐	☐
6	Do staff know the procedure for the reporting of complaints?	☐	☐	☐
7		☐	☐	☐
8		☐	☐	☐
9		☐	☐	☐
10		☐	☐	☐

Owner:

Assessment date:
Re-assessment date:

Topic	Hazard	Risk	People at Risk	Risk Rating	Control Measures	Actioned X	Residual Risk
Electrical fixed and portable appliances	Electric shock, burns etc	Personal injury	Contractors	Low/ med	• **Electrical Safety Certificates for circuits and fixed equipment should be retained on site following test and inspection of the electrical system** • **Electrical apparatus, both portable and fixed, to be checked and inspected periodically in accordance with organisational policy by a competent person** • **All work on live electrical equipment, eg distribution panels to be subject to a Permit to Work** • **A listing (inventory) of all portable electrical appliances such as drills, kettles, heaters and office equipment provided by the organisation, to be maintained** • **All portable appliances subject to inspection and test should be marked (label/sticker) as to their current status, eg date of next test etc** • **Portable electrical appliances to be visually inspected, prior to use by the user for signs of damage or fault, eg loose wiring, scorch marks, hot plugs etc** • **All users are to report electrical equipment which is poorly maintained** • **Suspect or faulty electrical apparatus must be taken out of use immediately, put in a secure place and labelled 'do not use' until attended to by a competent person**	☐ ☐ ☐ ☐ ☐ ☐ ☐ ☐ ☐ ☐	Low

Owner:

Assessment date:
Re-assessment date:

Questions the Risk Assessor needs to ask

No	Question	Yes	No	N/A
1	Are electrical fixed and portable appliances checked in accordance with regulation?	☐	☐	☐
2	Is it maintained and checked by competent and certificated electrical engineers?	☐	☐	☐
3	Are records of maintenance, checks and emergency repairs maintained current and available for inspection?	☐	☐	☐
4	Are staff aware of the emergency procedures to adopt in the event of defective and/or dangerous electrical equipment?	☐	☐	☐
5	Have all relevant risk assessments been completed and details given to staff affected?	☐	☐	☐
6	Is a log kept up to date for all portable electrical equipment in use?	☐	☐	☐
7	Are procedures in place to immediately withdraw faulty appliances and render them unusable?	☐	☐	☐
8		☐	☐	☐
9		☐	☐	☐
10		☐	☐	☐

Owner:

Assessment date:
Re-assessment date:

Topic	Hazard	Risk	People at Risk	Risk Rating	Control Measures	Actioned X	Residual Risk
Occupa-tional violence	Personal harm	Personal injury	Staff	Med/low	• **Developed procedures and mechanisms in place in the event of a violent incident**	☐	Low
					• **All staff trained as to requirements and on the developed procedures following full consultation with directly affected staff**	☐	
						☐	
						☐	
						☐	
						☐	
						☐	
						☐	
						☐	
						☐	

Owner:

Assessment date:
Re-assessment date:

Questions the Risk Assessor needs to ask

No	Question	Yes	No	N/A
1	Have the prevention of violence and associated procedures been agreed and adopted?	☐	☐	☐
2	Have risk assessments been carried out and staff considered to be at risk identified?	☐	☐	☐
3	Have control measures been actioned?	☐	☐	☐
4	Are reporting procedures active and monitored?	☐	☐	☐
5	Is training implemented and given by competent trainers?	☐	☐	☐
6	Are work activities monitored to ensure risks of violence are minimised?	☐	☐	☐
7	Are all incidents investigated and findings including actions required implemented?	☐	☐	☐
8		☐	☐	☐
9		☐	☐	☐
10		☐	☐	☐

Owner:

Assessment date:
Re-assessment date:

General

Topic	Hazard	Risk	People at Risk	Risk Rating	Control Measures	Actioned X	Residual Risk
Contractors – general	Property and personal	Premise and personal damage/ injury			• **Contractors are required to follow reporting instructions detailed in the order/contract (ie reporting to premise manager on arrival etc)** • **Contracts for work/services on the premises only to be left to contractors from relevant accredited lists or to contractors who have satisfied agreed company criteria** • **Contractors are required to be notified and to comply with organisation's relevant policies and procedures to ensure safety of all third parties** • **Contractors are required to wear ID badge at all times** • **Contractors are to be notified and to comply with any local office rules etc** • **Work is to be carried out giving due consideration to the staff and visitors at all times. Methods of work are to be discussed/recorded and agreed** • **All contractors should provide evidence to the organisation of adequate insurance cover**	☐ ☐ ☐ ☐ ☐ ☐ ☐ ☐ ☐ ☐	

Owner:

Assessment date:
Re-assessment date:

Questions the Risk Assessor needs to ask

No	Question	Yes	No	N/A
1	Are contractors' work activities monitored?	☐	☐	☐
2	Are the insurance aspects of contractors checked to ensure current status?	☐	☐	☐
3	Is the safety documentation for contractors who are on the organisation's accredited lists actively checked on at least a yearly basis to ensure current status?	☐	☐	☐
4	Are all contractors used required to wear ID badges at all times?	☐	☐	☐
5	Are permit-to-work activities and high-security activity supervised and monitored?	☐	☐	☐
6		☐	☐	☐
7		☐	☐	☐
8		☐	☐	☐
9		☐	☐	☐
10		☐	☐	☐

Owner:

Assessment date:
Re-assessment date:

Topic	Hazard	Risk	People at Risk	Risk Rating	Control Measures	Actioned X	Residual Risk
Cooling towers	Bacterial	Personal health affected	Staff and Passers-by	Low/ med	• **Ensure that experienced competent persons are maintaining the system in accordance with HSE and company guidelines** • **Check that there is a written schedule defiining maintenance, water treatment and testing** • **Check that the cooling tower(s) are registered with the local authority**	☐ ☐ ☐ ☐ ☐ ☐ ☐ ☐ ☐ ☐	Low

Owner:

Assessment date:
Re-assessment date:

Questions the Risk Assessor needs to ask

No	Question	Yes	No	N/A
1	Are cooling towers used by the organisation and have they been registered?	☐	☐	☐
2	Are the maintenance procedures current and applicable to codes of practice and current guidance?	☐	☐	☐
3	Is the maintenance carried out by competent contractors?	☐	☐	☐
4	Is the documentation for maintenance, repair and monitoring available for inspection and is it current?	☐	☐	☐
5		☐	☐	☐
6		☐	☐	☐
7		☐	☐	☐
8		☐	☐	☐
9		☐	☐	☐
10		☐	☐	☐

Owner:

Assessment date:
Re-assessment date:

Topic	Hazard	Risk	People at Risk	Risk Rating	Control Measures	Actioned X	Residual Risk
Tanks, taps and shower outlets	Bacterial Slips, trips and falls	Personal injury and health damage	Staff and visitors	Low	• **Ensure that a visual inspection is undertaken of all tanks and lagging by competent persons in accordance with organisational policy** • **All water storage tanks to be provided with covers to the top of the tank to prevent debris etc, from falling into the tank** • **Ensure that adequate access is provided to tank rooms, eg ladders.** • **Ensure that adequate lighting is provided to tank rooms** • **Overflow pipes to water storage distribution tanks should be screened to prevent ingress of rodents/vermin etc** • **Checks should also be made for leaks and corrosion etc** • **In shower rooms etc, where water usage is low – run outlet taps and shower heads for a minimum of 5 minutes (every 10 days) to reduce the risk of bacterial build up** • **Records of maintenance and inspections are to be maintained**	☐ ☐ ☐ ☐ ☐ ☐ ☐ ☐ ☐ ☐	Low

Owner:

Assessment date:
Re-assessment date:

Questions the Risk Assessor needs to ask

No	Question	Yes	No	N/A
1	**Are all water tanks inspected and checked for cleanliness and purity?**	☐	☐	☐
2	**Is the access to all tanks clear and unobstructed?**	☐	☐	☐
3	**Are ladders used for access and the lighting suitable, sound and free from defects?**	☐	☐	☐
4	**Are tanks lagged and overflow exit pipes suitably protected from access by vermin?**	☐	☐	☐
5	**Are all maintenance and repairs documented and maintained current and is it available for inspection?**	☐	☐	☐
6		☐	☐	☐
7		☐	☐	☐
8		☐	☐	☐
9		☐	☐	☐
10		☐	☐	☐

Owner:

Assessment date:
Re-assessment date:

Topic	Hazard	Risk	People at Risk	Risk Rating	Control Measures	Actioned X	Residual Risk
Hot and cold water systems	Bacterial Legionellae	Personal health	Staff and visitors	Low	• **Formal systems for water treatment and equipment maintenance are to be established on each building/premises that has hot and cold water**	☐	Low
					• **Up to date drawings of plant to be available**	☐	
					• **Description of system, including hot and cold supplies, to be identified**	☐	
					• **Guidance on normal operation easily available**	☐	
					• **Planned maintenance activities identified and easily obtained**	☐	
					• **Water quality checks to be in place and actioned for legionellae**	☐	
					• **Hot water should be stored at 60°C or above – checks to be periodically carried out in accordance with organisational policy**	☐	
					• **Check that the pipe runs do not have areas accessible to staff and others such that contact burns may occur**	☐	
					• **Check the pipe runs in plant rooms are marked as to contents as visual warnings to maintenance staff**	☐	
					• **Check hot water outlets for temperature and scale. Required temperature range 50–55°C after one minute**	☐	

Owner:

Assessment date:
Re-assessment date:

Questions the Risk Assessor needs to ask

No	Question	Yes	No	N/A
1	Is the hot and cold water supply regularly maintained by competent persons?	☐	☐	☐
2	Are schematic drawings available for use in the event of an emergency?	☐	☐	☐
3	Are water quality checks being maintained and documented by competent persons?	☐	☐	☐
4	Are all hot water pipes adequately protected to prevent contact burns?	☐	☐	☐
5	Are all hot water storage systems thermostatically controlled at the correct temperature?	☐	☐	☐
6		☐	☐	☐
7		☐	☐	☐
8		☐	☐	☐
9		☐	☐	☐
10		☐	☐	☐

Owner:

Assessment date:
Re-assessment date:

Topic	Hazard	Risk	People at Risk	Risk Rating	Control Measures	Actioned X	Residual Risk
Access to external roof areas	Slips, trips, falls and lockout	Personal injury	Staff and contractors	Med	• **Identify and inform staff/contractors of third-party equipment, ie microwave transmitting/receiving equipment** • **Work actioned or equipment positioned that prevents the use of trailing cables etc** • **Procedures to be in place to report any observed damage or bad working practices** • **Tools/equipment not to be left loose or unsecured on roof area** • **Secure access door/hatch when leaving the area**	☐ ☐ ☐ ☐ ☐ ☐ ☐ ☐ ☐ ☐	Low

Owner:

Assessment date:
Re-assessment date:

Questions the Risk Assessor needs to ask

No	Question	Yes	No	N/A
1	Are all access points to roof areas maintained secure at all times?	☐	☐	☐
2	If ladders are used to gain access are they in good condition and maintained/inspected regularly?	☐	☐	☐
3	Are there lone worker procedures in place to prevent isolation or lockout of persons?	☐	☐	☐
4		☐	☐	☐
5		☐	☐	☐
6		☐	☐	☐
7		☐	☐	☐
8		☐	☐	☐
9		☐	☐	☐
10		☐	☐	☐

Owner:

Assessment date:
Re-assessment date:

Roof Voids etc

Topic	Hazard	Risk	People at Risk	Risk Rating	Control Measures	Actioned X	Residual Risk
Internal access points – roof voids etc	Slips, trips and falls from height Manual handling	Personal injury	Staff Contractor	Low	• **Light switch to be accessible from outside the internal roof access point or immediately inside of the access** • **Purpose-designed loft ladder to be provided to enable safe access to loft area** • **Ladder to be subject to regular visual inspection to ensure that it remains in a serviceable condition** • **Ladder Safety Procedures to be in adopted at all times, ie climb ladder 'hands free', do not carry equipment etc** • **No smoking rule to be observed at all times** • **Adequate lighting to be available in loft area** • **Access panels to be locked at all times or access restricted by position to unauthorised persons when not in use** • **Access points near/over stairwells to be suitably guarded to prevent falls down into stairwell** • **Roof edges to be protected or access restricted whilst persons are working from the roof areas**	☐ ☐ ☐ ☐ ☐ ☐ ☐ ☐ ☐ ☐	Low

Owner:

Assessment date:
Re-assessment date:

Questions the Risk Assessor needs to ask

No	Question	Yes	No	N/A
1	Are access panels to roof voids maintained secure at all times?	☐	☐	☐
2	Are ladders used for any access maintained free from defects and regularly inspected?	☐	☐	☐
3	Is the lighting provided suitable and sufficient?	☐	☐	☐
4	Are the light switches accessible externally or easily from just inside the void area?	☐	☐	☐
5	Are no smoking signs posted in the void areas?	☐	☐	☐
6	Are the voids free from obstruction and in appropriate or unsafe storage items?	☐	☐	☐
7		☐	☐	☐
8		☐	☐	☐
9		☐	☐	☐
10		☐	☐	☐

Owner:

Assessment date:
Re-assessment date:

Window Cleaning

External and Internal

Topic	Hazard	Risk	People at Risk	Risk Rating	Control Measures	Actioned X	Residual Risk
Window cleaning – external and internal	Falls, falling materials	Personal injury	Members of public/ staff	Low	• **Where windows are to be cleaned from the inside, and are above 1st floor level (6m), provision to be made for the windows to pivot** • **Where windows are to be cleaned externally from ladders and are above 6m, ladders must be tied and a solid level footing available. Eye bolts to be used where fitted** • **If eyebolts are provided on the face/internally of the building then they must be placed on an inspection and test programme in accordance with British Standards** • **Where windows are to be cleaned externally and are above 9m ladder access is prohibited and alternative means of access is required, eg tower scaffold** • **Only competent and trained persons to use abseiling or personal suspension equipment whilst cleaning**	☐ ☐ ☐ ☐ ☐ ☐ ☐ ☐ ☐ ☐	Low

Owner:

Assessment date:
Re-assessment date:

Questions the Risk Assessor needs to ask

No	Question	Yes	No	N/A
1	Are suitable anchorage points sited for window cleaner's use?	☐	☐	☐
2	Are the anchorages tested and maintained free from damage?	☐	☐	☐
3	Are external eyebolts similarly tested and maintained by competent persons to British Standards and/or organisation's insurer's requirements?	☐	☐	☐
4	Have window cleaning safe working procedures been previously agreed with contractor and are they current to present needs of the organisation?	☐	☐	☐
5	Have risk assessments been completed and submitted by the contractor and are they suitable and sufficient?	☐	☐	☐
6		☐	☐	☐
7		☐	☐	☐
8		☐	☐	☐
9		☐	☐	☐
10		☐	☐	☐

Owner:

Assessment date:
Re-assessment date:

Use of Cradles

Topic	Hazard	Risk	People at Risk	Risk Rating	Control Measures	Actioned X	Residual Risk
Window cleaning – use of cradles	Falls, falling materials	Personal injury			• **Where windows are to be cleaned from access cradles the following is required:** ○ Equipment to be safely and securely stored when not in use ○ Equipment to be subject to routine inspection and maintenance for load-bearing equipment ○ Window-cleaning contractors to be assessed for suitability, including taking up references, reviewing safety policy etc ○ Window-cleaning contractors to provide details on risk assessment and safe method of work ○ Window-cleaning employees to be adequately trained and supervised in the use of the equipment	☐ ☐ ☐ ☐ ☐ ☐ ☐ ☐ ☐ ☐	

Owner:

Assessment date:
Re-assessment date:

Questions the Risk Assessor needs to ask

No	Question	Yes	No	N/A
1	Are window-cleaning cradles used and are they used by trained contractors?	☐	☐	☐
2	Have safe methods of work been agreed prior to the start of the work or contract?	☐	☐	☐
3	Are the cradles maintained by a competent and trained person or organisation?	☐	☐	☐
4	Has the window-cleaning contractor supplied full and sufficient risk assessments?	☐	☐	☐
5	Can the contractor confirm officially that their employees have been suitably trained and are adequately supervised whilst using cradles?	☐	☐	☐
6		☐	☐	☐
7		☐	☐	☐
8		☐	☐	☐
9		☐	☐	☐
10		☐	☐	☐

Owner:

Assessment date:
Re-assessment date:

General Office

Topic	Hazard	Risk	People at Risk	Risk Rating	Control Measures	Actioned X	Residual Risk
Work equipment – general office	Physical damage	Personal injury and health affected	Staff	Low	• **All equipment is to be designed and suitable for its intended use**	☐	Low
					• **Equipment to be subject to regular maintenance and records kept up to date**	☐	
					• **Ensure that staff using equipment have been trained as necessary**	☐	
					• **Ensure that staff maintaining the equipment have been trained as necessary**	☐	
					• **All dangerous parts of machinery, eg wheels and belts, hot surfaces etc to be guarded**	☐	
					• **Equipment provided with controls, including emergency stops, as necessary for safety**	☐	
					• **Ensure that equipment can be isolated from energy supply for the purposes of maintenance**	☐	
					• **Equipment properly installed and in a stable condition**	☐	
					• **Environment is suitable – well lit and ventilated**	☐	
					• **Equipment marked with appropriate information and warnings for safe use**	☐	

Owner:

Assessment date:
Re-assessment date:

Questions the Risk Assessor needs to ask

No	Question	Yes	No	N/A
1	**Is the general work equipment maintained safe by a competent and trained person?**	☐	☐	☐
2	**Is the maintenance records kept current and available for inspection?**	☐	☐	☐
3	**Has any equipment with moving parts been suitably guarded?**	☐	☐	☐
4	**Are staff trained in the use of the equipment?**	☐	☐	☐
5	**Is the training documented, maintained to current standards and is it available for inspection?**	☐	☐	☐
6	**Are the operating manuals available for staff to use?**	☐	☐	☐
7		☐	☐	☐
8		☐	☐	☐
9		☐	☐	☐
10		☐	☐	☐

Owner:

Assessment date:
Re-assessment date:

Topic	Hazard	Risk	People at Risk	Risk Rating	Control Measures	Actioned X	Residual Risk
Emergency lighting	Entrapment/ burns etc	Property and personal injury	Staff and Visitors	Low	• **Emergency lighting to be subject to periodic testing carried out by contractors under a service agreement in accordance with organisational policy, eg six monthly** • **Indicator lights (where fitted) on emergency lighting units to be checked regularly to ensure unit is operational** • **If indicator light is not illuminated then unit should be considered faulty and be reported immediately to relevant manager** • **Records to be maintained of all maintenance, inspections and tests of emergency lighting**	☐ ☐ ☐ ☐ ☐ ☐ ☐ ☐ ☐ ☐	Low

Owner:

Assessment date:
Re-assessment date:

Questions the Risk Assessor needs to ask

No	Question	Yes	No	N/A
1	**If fitted, is the emergency lighting suitably maintained and regularly checked by a competent person or contractor?**	☐	☐	☐
2	**Are records maintained as to repair and maintenance?**	☐	☐	☐
3	**Are the records available for inspection?**	☐	☐	☐
4	**Is the emergency lighting sufficient for current premises' use?**	☐	☐	☐
5		☐	☐	☐
6		☐	☐	☐
7		☐	☐	☐
8		☐	☐	☐
9		☐	☐	☐
10		☐	☐	☐

Owner:

Assessment date:
Re-assessment date:

General Fire Precautions

Topic	Hazard	Risk	People at Risk	Risk Rating	Control Measures	Actioned X	Residual Risk
General fire precautions	Fire and fire spread and emergency escape	Property and personal injury	Staff and Visitors	Low	• **Non essential electrical appliances switched off and unplugged when not in use** • **Plugs removed carefully, no pulling by the flex** • **Staff aware of the warning signs of dangerous wiring:** ○ brown hot plugs and sockets ○ fuses that blow for no obvious reason ○ lights flickering ○ scorch marks on sockets and plugs • **Any faults discovered and reported must only be repaired by competent trained contractor** • **No smoking policy for the premises implemented and monitored** • **Ensure that regular housekeeping maintains all areas, eg cupboards, waste bins etc, free from excessive materials, which may fuel fires** • **Ensure that all hazardous materials, eg flammable products are stored safely** • **When contractors are employed to work in the premises ensure that adequate provision has been made to control 'Hot Works', eg plumbing**	☐ ☐ ☐ ☐ ☐ ☐ ☐ ☐ ☐ ☐	Low

Owner:

Assessment date:
Re-assessment date:

Questions the Risk Assessor needs to ask

No	Question	Yes	No	N/A
1	Are staff aware of general fire safety procedures and, in particular, the requirements to maintain fire doors in a closed position at all times unless linked via electromagnetic systems?	☐	☐	☐
2	Is there a requirement for staff to be given refresher fire safety training?	☐	☐	☐
3	Are new staff given all of the fire safety details on induction?	☐	☐	☐
4	Is a no smoking policy in effect and is it monitored and enforced?	☐	☐	☐
5	Are housekeeping standards monitored and maintained at a high level?	☐	☐	☐
6		☐	☐	☐
7		☐	☐	☐
8		☐	☐	☐
9		☐	☐	☐
10		☐	☐	☐

Owner:

Assessment date:
Re-assessment date:

Fire Extinguishers

Topic	Hazard	Risk	People at Risk	Risk Rating	Control Measures	Actioned X	Residual Risk
Fire extin-guishers	Personal and corporate damage	Property and personal injury	Staff and visitors	Low	• **Fire extinguishers to be subject to an annual test and inspection carried out by contractors under a service agreement** • **Regular visual inspection to be made on fire extinguishers** • **Any non-conformances to be reported to the service contractor immediately** • **All fire extinguishers points to remain free from obstructions at all times** • **Records to be maintained by the relevant site manager of tests and inspections (individual fire extinguishers will be marked at the time of inspection)**	☐ ☐ ☐ ☐ ☐ ☐ ☐ ☐ ☐ ☐	Low

Owner:

Assessment date:
Re-assessment date:

Questions the Risk Assessor needs to ask

No	Question	Yes	No	N/A
1	**Are the fire extinguishers maintained by a competent contractor or person?**	☐	☐	☐
2	**Are they suitably marked to show the last date of check or maintenance?**	☐	☐	☐
3	**Are the extinguishers suitably marked with instructions as to use?**	☐	☐	☐
4	**Are all the fire extinguisher areas maintained free from obstruction?**	☐	☐	☐
5	**Are the test and maintenance records available for inspection?**	☐	☐	☐
6		☐	☐	☐
7		☐	☐	☐
8		☐	☐	☐
9		☐	☐	☐
10		☐	☐	☐

Owner:

Assessment date:
Re-assessment date:

Fire Alarm System/ Break Glass Points

Topic	Hazard	Risk	People at Risk	Risk Rating	Control Measures	Actioned X	Residual Risk
Fire alarm system and break glass points	Fire, entrapment	Property and personal injury	Staff and visitors	Low	• **Fire alarm system, including battery back up to be subject to an annual test by contractors under a service agreement** • **A weekly test actioned by activating a different call point of the fire alarm system to ensure that the system remains fully operational** • **Any defects/faults to be recorded and action taken to rectify them** • **Records to be maintained of all annual and weekly tests of the fire alarm system**	☐ ☐ ☐ ☐ ☐ ☐ ☐ ☐ ☐ ☐	Low

Owner:

Assessment date:
Re-assessment date:

Questions the Risk Assessor needs to ask

No	Question	Yes	No	N/A
1	Are the fire alarm call points maintained by a suitably trained and competent person or contractor?	☐	☐	☐
2	Are records kept of call point tests?	☐	☐	☐
3	Are defects reported and repaired without delay?	☐	☐	☐
4	Whilst awaiting repair of any defective fire alarm call point, are staff and fire marshals informed of estimated repair time?	☐	☐	☐
5		☐	☐	☐
6		☐	☐	☐
7		☐	☐	☐
8		☐	☐	☐
9		☐	☐	☐
10		☐	☐	☐

Owner:

Assessment date:
Re-assessment date:

Fire Alarm – Panel Indicator

Topic	Hazard	Risk	People at Risk	Risk Rating	Control Measures	Actioned X	Residual Risk
Fire alarm panel indicator	Fire and non-notification	Property and personal injury	Staff and visitors	Low	• **The fire alarm panel indicator should be checked weekly for normal operating condition** • **Faults reported immediately to the contractor** • **Records maintained of all maintenance and inspections carried out on the main control panel**	☐ ☐ ☐ ☐ ☐ ☐ ☐ ☐ ☐ ☐	Low

Owner:

Assessment date:
Re-assessment date:

Questions the Risk Assessor needs to ask

No	Question	Yes	No	N/A
1	Is the fire alarm panel tested and maintained by a competent and trained contractor or person?	☐	☐	☐
2	Are defects or breakdowns immediately reported?	☐	☐	☐
3	Are records maintained of repairs and inspections?	☐	☐	☐
4	Are the written records available for inspection?	☐	☐	☐
5		☐	☐	☐
6		☐	☐	☐
7		☐	☐	☐
8		☐	☐	☐
9		☐	☐	☐
10		☐	☐	☐

Owner:

Assessment date:
Re-assessment date:

Electro-magnetic Fire Doors

Topic	Hazard	Risk	People at Risk	Risk Rating	Control Measures	Actioned X	Residual Risk
Electro-magnetic fire doors	Fire and fire spread	Property and personal injury	Staff and visitors	Low	• **Electro-magnetic door holders checked periodically by operating the manual override button to ensure that the door closes properly against doorframe** • **All doors activated during the weekly test of the fire alarm – any doors not closing automatically to be attended/repaired immediately** • **Records maintained of all maintenance and service checks**	☐ ☐ ☐ ☐ ☐ ☐ ☐ ☐ ☐ ☐	Low

Owner:

Assessment date:
Re-assessment date:

Questions the Risk Assessor needs to ask

No	Question	Yes	No	N/A
1	Are electro-magnetic doors maintained free from defects by a competent and trained person or contractor?	☐	☐	☐
2	Are records maintained of repairs and inspections?	☐	☐	☐
3	Are the weekly tests included in the records?	☐	☐	☐
4		☐	☐	☐
5		☐	☐	☐
6		☐	☐	☐
7		☐	☐	☐
8		☐	☐	☐
9		☐	☐	☐
10		☐	☐	☐

Owner:

Assessment date:
Re-assessment date:

Internal Fire Doors and Closure Mechanisms

Topic	Hazard	Risk	People at Risk	Risk Rating	Control Measures	Actioned X	Residual Risk
Internal fire doors and closure mecha-nisms	Fire and fire spread and emergency escape	Property and personal injury	Staff and visitors	Low	• **All corridor doors monitored etc to ensure that they remain easy to operate and will not cause any person who may be frail or disabled any problem** • **Carpets/tiles etc inspected to ensure that they do not restrict door from opening or closing** • **Fire doors not propped open** • **Fire door smoke seals, situated around door or doorframe inspected and remain free from paint etc** • **Records maintained of all maintenance and service checks** • **Doors failing to shut correctly repaired as soon as possible** • **Door closure mechanisms visually checked periodically for signs of leaking oil – renewed or repaired as necessary** • **Speed of door closures monitored – if too fast then damage will be caused to doorframe, surrounding wall or a member of staff or visitors**	☐ ☐ ☐ ☐ ☐ ☐ ☐ ☐ ☐ ☐	Low

Owner:

Assessment date:
Re-assessment date:

Questions the Risk Assessor needs to ask

No	Question	Yes	No	N/A
1	Are the fire door closure mechanisms maintained free from defect?	☐	☐	☐
2	Are checks made on the mechanisms to ensure free movement over the flooring?	☐	☐	☐
3	Are written records maintained of inspections and repairs?	☐	☐	☐
4	Are the records available for inspection?	☐	☐	☐
5	Are fire doors being propped open?	☐	☐	☐
6		☐	☐	☐
7		☐	☐	☐
8		☐	☐	☐
9		☐	☐	☐
10		☐	☐	☐

Owner:

Assessment date:
Re-assessment date:

External Fire Doors

Topic	Hazard	Risk	People at Risk	Risk Rating	Control Measures	Actioned X	Residual Risk
External fire doors	Fire and fire spread and escape restrictions	Property and personal injury	Staff and visitors	Low	• **Fire doors leading to the outside of the building suitably signed and remain free from obstruction at all times**	☐	Low
					• **Checks made periodically of all doors to ensure that they are operational in accordance with organisational policy**	☐	
					• **The outside of the fire exit door signed 'Fire door – do not obstruct'**	☐	
					• **Records maintained of all maintenance and service checks**	☐	
						☐	
						☐	
						☐	
						☐	
						☐	
						☐	

Owner:

Assessment date:
Re-assessment date:

Questions the Risk Assessor needs to ask

No	Question	Yes	No	N/A
1	**Are all external exiting fire doors suitably signed?**	☐	☐	☐
2	**Are the doors regularly tested and maintained?**	☐	☐	☐
3	**Are the doors maintained free from obstructions, internally and externally, at all times?**	☐	☐	☐
4	**Are the doors suitably marked and signed externally?**	☐	☐	☐
5	**Are written records maintained of repairs and maintenance including tests?**	☐	☐	☐
6	**Are the written records available for inspection?**	☐	☐	☐
7		☐	☐	☐
8		☐	☐	☐
9		☐	☐	☐
10		☐	☐	☐

Owner:

Assessment date:
Re-assessment date:

Premises Evacuation

Topic	Hazard	Risk	People at Risk	Risk Rating	Control Measures	Actioned X	Residual Risk
Premise evacuation (where applicable)	Fire and fire spread and emergency escape	Property and personal injury	Staff and visitors	Low	• **Fire routines and procedures in place regarding premise evacuation**	☐	Low
					• **Nominated fire wardens appointed and suitably trained**	☐	
					• **In an emergency, assess the situation and call the emergency services in line with written procedures**	☐	
					• **Fire routines and procedures, once finalised, made known to all staff**	☐	
					• **Full scale fire drills, eg evacuation carried out in accordance with the fire certificate or fire risk assessment**	☐	
					• **Fire plans and risk assessments in place and addresses the following:** ○ sounding the alarm ○ safe evacuation – if required ○ caring for injured persons ○ fighting the fire – using fire extinguishers ○ identification of master service control switches – gas and electricity	☐	
						☐	
						☐	

Owner:

Assessment date:
Re-assessment date:

Questions the Risk Assessor needs to ask

No	Question	Yes	No	N/A
1	Are fire routine notices suitably fixed around the work area and at exit points?	☐	☐	☐
2	Are the emergency fire evacuation plans understood by all staff?	☐	☐	☐
3	Are there assembly points identified and suitably marked?	☐	☐	☐
4	Are fire wardens suitably positioned throughout the organisation?	☐	☐	☐
5	Are staff aware of who the fire wardens are and their role in an evacuation?	☐	☐	☐
6	Are fire drills actioned and recorded?	☐	☐	☐
7	Are the records available for inspection?	☐	☐	☐
8	Do the reception staff, security teams if applicable and all staff know the procedure for reporting an incident?	☐	☐	☐
9	Have all special fire risks been brought to the attention of staff and have they been informed of the procedures and actions to adopt in the event of an incident involving the special risk?	☐	☐	☐
10		☐	☐	☐

Owner:

Assessment date:
Re-assessment date:

External Fire Stairs

Topic	Hazard	Risk	People at Risk	Risk Rating	Control Measures	Actioned X	Residual Risk
External fire stairs	Fire and fire spread and emergency escape	Property and personal injury	Staff and visitors	Low	• **External fire escape stairs to have slip resistant surface, suitable handrails and provided with weatherproof emergency lighting** • **External fire escape stairs maintained in a safe condition, eg structural stability, free from excessive rust etc**	☐ ☐ ☐ ☐ ☐ ☐ ☐ ☐ ☐ ☐	Low

Owner:

Assessment date:
Re-assessment date:

Questions the Risk Assessor needs to ask

No	Question	Yes	No	N/A
1	Are external fire escapes maintained free from obstruction and slip hazards?	☐	☐	☐
2	Are the external fire escapes property checked and inspected/tested by competent contractors?	☐	☐	☐
3	Are records kept of maintenance and any tests and are they available for inspection?	☐	☐	☐
4		☐	☐	☐
5		☐	☐	☐
6		☐	☐	☐
7		☐	☐	☐
8		☐	☐	☐
9		☐	☐	☐
10		☐	☐	☐

Owner:

Assessment date:
Re-assessment date:

Fire Signs

Topic	Hazard	Risk	People at Risk	Risk Rating	Control Measures	Actioned X	Residual Risk
Fire signs	Emergency escape	Property and personal injury	Staff and visitors	Low	• **Designated fire exit routes adequately signed in accordance with current British Standards, ie 'running person'** • **Where illuminated signs are fitted they are to be subject to operational tests in accordance with British Standards**	☐ ☐ ☐ ☐ ☐ ☐ ☐ ☐ ☐ ☐	Low

Owner:

Assessment date:
Re-assessment date:

Questions the Risk Assessor needs to ask

No	Question	Yes	No	N/A
1	Are fire signs maintained in accordance with the current regulations?	☐	☐	☐
2	Are the signs maintained clean and free from obstruction?	☐	☐	☐
3	Are written records maintained regarding repairs and tests?	☐	☐	☐
4	Are the records available for inspection?	☐	☐	☐
5		☐	☐	☐
6		☐	☐	☐
7		☐	☐	☐
8		☐	☐	☐
9		☐	☐	☐
10		☐	☐	☐

Owner:

Assessment date:
Re-assessment date:

ACTION PLAN

SECTION:

SITE:

RISK ASSESSMENT ITEM NO	ACTION REQUIRED	ACTION BY	PRIORITY 1-5	TARGET DATE	COST	COMPLETED (Name/Signature)
-			-			
-			-			
-			-			
-			-			
-			-			
-			-			
-			-			
-			-			

KEY
Priority
1
2
3
4
5

Action Plan Prepared by (name):

Signature:

Date:

ACTION PLAN

SECTION:

SITE:

RISK ASSESSMENT ITEM NO	ACTION REQUIRED	ACTION BY	PRIORITY 1-5	TARGET DATE	COST	COMPLETED (Name/Signature)
-			-			
-			-			
-			-			
-			-			
-			-			
-			-			
-			-			
-			-			

KEY
Priority
1
2
3
4
5

Action Plan Prepared by (name):

Signature:

Date:

ACTION PLAN

SECTION:

SITE:

RISK ASSESSMENT ITEM NO	ACTION REQUIRED	ACTION BY	PRIORITY 1-5	TARGET DATE	COST	COMPLETED (Name/Signature)
-			-			
-			-			
-			-			
-			-			
-			-			
-			-			
-			-			
-			-			

KEY
Priority
1
2
3
4
5

Action Plan Prepared by (name):

Signature:

Date:

ACTION PLAN

SECTION:

SITE:

RISK ASSESSMENT ITEM NO	ACTION REQUIRED	ACTION BY	PRIORITY 1-5	TARGET DATE	COST	COMPLETED (Name/Signature)
-			-			
-			-			
-			-			
-			-			
-			-			
-			-			
-			-			
-			-			

KEY
Priority
1
2
3
4
5

Action Plan Prepared by (name):

Signature:

Date:

ACTION PLAN

SECTION:

SITE:

RISK ASSESSMENT ITEM NO	ACTION REQUIRED	ACTION BY	PRIORITY 1-5	TARGET DATE	COST	COMPLETED (Name/Signature)
-			-			
-			-			
-			-			
-			-			
-			-			
-			-			
-			-			
-			-			

KEY
Priority
1
2
3
4
5

Action Plan Prepared by (name):

Signature:

Date:

ACTION PLAN

SECTION:

SITE:

RISK ASSESSMENT ITEM NO	ACTION REQUIRED	ACTION BY	PRIORITY 1-5	TARGET DATE	COST	COMPLETED (Name/Signature)
-			-			
-			-			
-			-			
-			-			
-			-			
-			-			
-			-			
-			-			

KEY
Priority
1
2
3
4
5

Action Plan Prepared by (name):

Signature:

Date:

ACTION PLAN

SECTION:

SITE:

RISK ASSESSMENT ITEM NO	ACTION REQUIRED	ACTION BY	PRIORITY 1-5	TARGET DATE	COST	COMPLETED (Name/Signature)
-			-			
-			-			
-			-			
-			-			
-			-			
-			-			
-			-			
-			-			

KEY
Priority
1
2
3
4
5

Action Plan Prepared by (name):

Signature:

Date:

ACTION PLAN

SECTION:

SITE:

RISK ASSESSMENT ITEM NO	ACTION REQUIRED	ACTION BY	PRIORITY 1-5	TARGET DATE	COST	COMPLETED (Name/Signature)
-			-			
-			-			
-			-			
-			-			
-			-			
-			-			
-			-			
-			-			

KEY
Priority
1
2
3
4
5

Action Plan Prepared by (name):

Signature:

Date: